STUDY GUIDE

Nancy J. Obermeyer

Indiana State University

Diversity Amid Globalization

WORLD REGIONS, ENVIRONMENT, DEVELOPMENT

Fifth Edition

Les Rowntree

Martin Lewis

Marie Price

William Wyckoff

Prentice Hall

Boston   Columbus   Indianapolis   New York   San Francisco   Upper Saddle River
Amsterdam   Cape Town   Dubai   London   Madrid   Milan   Munich   Paris   Montréal   Toronto
Delhi   Mexico City   São Paulo   Sydney   Hong Kong   Seoul   Singapore   Taipei   Tokyo

Geography Editor: Christian Botting
Marketing Manager: Maureen McLaughlin
Assistant Editor: Kristen Sanchez
Editorial Assistant: Bethany Sexton
Marketing Assistant: Nicola Houston
Managing Editor, Geosciences and Chemistry: Gina M. Cheselka
Project Manager, Science: Wendy A. Perez
Operations Specialist: Maura Zaldivar
Supplement Cover Designer: Paul Gourhan
Cover Photograph: Landscape with grain terraces, Highlands, Madagascar, Africa: Alamy

The author and publisher of this book have used their best efforts in preparing this book. These efforts include the development, research, and testing of the theories and programs to determine their effectiveness. The author and publisher make no warranty of any kind, expressed or implied, with regard to these programs or the documentation contained in this book. The author and publisher shall not be liable in any event for incidental or consequential damages in connection with, or arising out of, the furnishing, performance, or use of these programs.

Printed in the United States of America

10   9   8   7   6   5   4   3   2   1

ISBN-13:   978-0-321-75091-4
ISBN-10:   0-321-75091-8

Prentice Hall
is an imprint of

www.pearsonhighered.com

Contents

Chapter 1

Diversity Amid Globalization

Learning Objectives

➢ Globalization, defined as the increasing interconnectedness of people and places through converging economic, political, and cultural activities, is introduced as the key concept for the book
➢ This chapter provides an overview of the key themes, including environment, population and settlement, cultural, geopolitics, and economic concepts, and measures that are used in subsequent chapters
➢ This chapter introduces broad debates surrounding globalization as a concept and as a process
➢ This chapter introduces tension between globalization and diversity; what it means; and how to understand from a geographic point-of-view

Chapter Outline

1. **Converging Currents of Globalization**
    A. **Globalization and Cultural Change:** Globalization is accompanied by the spread of a global consumer culture; increasingly, Western values and culture are spread through media outlets
    B. **Globalization and Geopolitics:** Globalizing tendencies have transformed our understanding of national borders and the role of the nation-state
    C. **Environmental Concerns:** Globalization contributes to the creation and intensification of environmental problems; globalization might also promote global cooperation to respond to environmental threats
    D. **Social Dimensions:** Globalization has contributed to a greater integration of the world's population through migration; globalization also contributes to the expansion of criminal activities, including drug trafficking and prostitution
2. **Advocates and Critics of Globalization**
    A. **The Pro-Globalization Stance:** globalization is a logical and inevitable expression of contemporary capitalism; globalization benefits all nations and peoples through greater economic integration
    B. **Critics of Globalization:** globalization is not natural, but instead the product of unfair and unequal economic policies that favor wealthy countries over poorer countries; globalization promotes free-market, export-oriented economies at the expense of localized, sustainable activities
    C. **A Middle Position:** through changing technologies (e.g., the Internet) globalization is perhaps inevitable; globalization however can be managed to reduce economic inequalities
3. **Diversity in a Globalizing World**
    A. This section questions whether globalization is leading to a culturally homogenous world, or whether (and to what extent) the world remains a highly diverse place with local differences; a 'politics of diversity' is introduced, noting that people around the world are both actively promoting and resisting globalizing practices
4. **Geography Matters: Environments, Regions, Landscapes**
    A. **Geography and Geographic Inquiry:** Geography refers to the description and explanation of patterns and processes across Earth's surface; a fundamental component of geographic

inquiry is the examination of basic yet highly complex relationships between humans and their environments

    B. **Areal Differentiation and Integration:** the description and explanation of differences across a region or area

    C. **Regions:** units of similarity (spatial categories); few regions are completely homogenous; boundaries between regions are subjective and artificial

    D. **The Cultural Landscape:** humans transform space into distinct places, understood as cultural landscapes; these include the visible, material expression of human settlements and reflects the basic human needs of shelter, food, and work

    E. **Scale:** the size or geographic extent of an area; common scales include local, regional, and global levels of analysis

5. **Themes and Issues in World Regional Geography**

    A. **Environmental Geography:** refers to a region's climate, geology, hydrology, and vegetation

    B. **Population and Settlement:** refers to the size and distribution of Earth's human population

        a. **Population Growth and Change**

            i. **Rate of Natural Increase (RNI):** A statistic that expresses a country's or region's annual growth rate; expressed as a percentage

            ii. **Total Fertility Rate:** a synthetic rate that expresses the fertility of a country or region; commonly understood as the average number of children a woman would give birth to during her reproductive years

            iii. **Young and Old Populations:** the proportion of young persons (i.e., those under 15 years of age) compared to the proportion of persons over 65 years of age; graphically illustrated by population pyramids, which also reflect sex differences of the population

            iv. **Life Expectancy:** the average number of years a person is expected to live from birth; influenced by health services, nutrition, and sanitation; used as an indicator of social development

            v. **The Demographic Transition:** a conceptual model that tracks changes in birth and death rates for a country or region over time

        b. **Migration Patterns**

            i. **Push and Pull Forces:** conditions that influence whether people move or stay; push forces include negative conditions, such as civil strife, environmental degradation, or unemployment, while pull forces include positive conditions, such as better economic opportunities and climates

            ii. **Net Migration Rate:** a statistic that depicts whether more people are entering or leaving a country or region

        c. **Settlement Geography**

            i. **An Urban World**

                1. **Urbanized population:** the percentage of a country's population living in cities; approximately half of the world's population currently lives in cities

            ii. **Conceptualizing the City**

                1. **Urban primacy:** describes a city that is disproportionately large and dominates economic, political, and cultural activities within a country; cities are termed 'primate cities'

                2. **Urban structure:** a term that refers to the distribution and patterning of land use within a city

                3. **Urban form:** describes the physical arrangement of buildings, streets, parks, and architecture; provides cities with a unique sense of place

                4. **Overurbanization:** a process of rapid urban growth; existing infrastructure (e.g., housing, transportation, waste disposal, and

2

water supply) is unable to meet the demands of the growing urban population

     5. **Squatter settlements:** illegal developments of makeshift housing on land neither owned nor rented by their inhabitants

C. **Cultural Coherence and Diversity:** The Geography of Tradition and Change

    a. **Culture in a Globalizing World**: culture is a learned and shared behavior; culture includes both abstract and material dimensions; globalization processes highlight that cultures are dynamic, not static

    b. **When Cultures Collide**

      i. **Cultural imperialism:** the active promotion of one cultural system at the expense of another

      ii. **Cultural nationalism:** the process of protecting and defending a cultural system

      iii. **Cultural syncretization or hybridization:** the blending of cultures to form a new type of culture

D. **Language and Culture in Global Context**

    a. Languages and cultures are intertwined

    b. Languages are classified into families, branches, and groups

    c. Dialect: a distinctive form associated with a specific language (e.g., American and British English)

    d. Lingua franca: a third language that is adopted by people from different cultural groups who cannot speak each other's language (e.g., Swahili is a lingua franca of Africa)

E. **A Geography of World Religions**

    a. Religions are important in defining cultural identity

    b. Universalizing religions attempt to appeal to all peoples and actively seek new converts (e.g., Christianity, Islam)

    c. Ethnic religions remain closely identified with specific groups of people; these usually do not actively seek new converts (e.g., Judaism, Hinduism)

    d. Secularization: a situation whereby people consider themselves either non-religious or atheistic

F. **Geopolitical Framework**

    a. Geopolitics refers to the relationship between geography and politics; focuses on the relations between power, territory, and space; usually emphasizes state-to-state relations

    b. **States:** a political entity with territorial boundaries recognized by other countries and internally governed

    c. **Nation:** a large group of people who share numerous socio-cultural elements, such as language, religion, tradition, and identity

    d. **Nation-state:** relatively homogenous cultural group occupying its own independent political territory

    e. **Colonialism:** the formal establishment of rule over a foreign population

    f. **Decolonialization:** the process of a colony gaining control over its own territory and establishing an independent government

    g. **Global terrorism and insurgency:** processes whereby rebellious or separatist groups seek independence, autonomy, and territorial control; terrorism refers to violence directed at non-military targets to achieve political goals

      i. Global terrorism is understood as both a product and a reaction to globalization

G. **Economic and Social Development**
   a. Economic development refers to increased prosperity to people, regions, and nations; it usually includes social improvements such as improved education systems, better health care, and progressive labor practices
   b. Economic development often exhibits a geographic unevenness of prosperity and social improvement; approximately half of the world's population is classified as living in poverty
   c. **More- and Less-Developed Countries:** refers to uneven levels of development between countries; usually expressed by the core-periphery model whereby economically 'advanced' countries such as the United States, Japan, England are seen as comprising the core, while economically marginal countries constitute the periphery; other terms used to refer to global economic inequalities include 'First' and 'Third' worlds and 'north–south' differences; the terms 'more-developed countries' (MDC) and 'less-developed countries' (LDC) are forwarded as more appropriate concepts
   d. **Indicators of Economic Development**
      i. **Development:** has both qualitative and quantitative dimensions; measures structural changes, such as shift from agricultural to manufacturing activities and changes in the allocation of labor, capital, and technology; development implies an improvement in standard of living, education, and political organization
      ii. **Growth:** refers to the increase in the size of system (e.g., agricultural or industrial output); economic systems may grow yet not lead to increased levels of development
   e. **Measurements of Growth and Development**
      i. **Gross Domestic Product (GDP):** a traditional measure of the size of a country's economy; it refers to the value of all final goods and services produced within a country's borders
      ii. **Gross National Income (GNI):** formerly known as the gross national product (GNP), this refers to the GDP plus net income from abroad; it is considered misleading and incomplete because it does not consider nonmarket economic activities; nor does it consider the degradation or depletion of natural resources that may limit future economic growth
      iii. **GNI per capita:** a country's GNI divided by its population; allows for country-to-country comparisons to be made despite vastly different sizes in population between countries
      iv. **Purchasing Power Parity (PPP):** a concept that provides a sense of the local cost of living that takes into consideration the strength or weakness of local currencies
      v. **Economic Growth Rates:** the average annual percent growth of a country's GDP
   f. **Indicators of Social Development:** measures that relate to the conditions and quality of human life
      i. **Human Development Index (HDI):** an index that combines data on life expectancy, literacy, educational attainment, gender equity, and income; conveys a sense of a country's human and social development
      ii. **Poverty:** the international definition of poverty is living on less than U.S. $2 per day; deep poverty is defined as living on less than U.S. $1 per day
      iii. **Under Age Five Mortality:** a widely used indicator of social conditions, the measurement refers to the number of children who die before five years of age per 1,000 people within that age bracket; mortality levels are influenced by availability of and access to food, health services, and public sanitation

4

iv. **Gender Equity:** refers to the different conditions and quality of life between men and women; a common measurement is the ratio of male to female students enrolled in primary and secondary schools

Summary

- Globalization is driving a fundamental reorganization of cultures and economies through changes in communication technologies, transnational firms, and Western consumer habits
- Globalization involves both positive and negative changes; it is controversial
- In many developing regions, population and settlement issues revolve around rapid population growth, family planning, migration, and rapid urbanization
- The trends of global cultural homogenization and the counter-currents of local cultural and ethnic identity result in considerable tension
- Terrorism has emerged as a primary global geopolitical issue
- The increasing disparity between the rich and the poor is a major economic and social issue of globalization

Research or Term Paper Ideas

- Compare and contrast the colonial experiences of countries in two different regions. For example, you might select Brazil and Angola, both of which were colonies of Portugal. When did the colonizers arrive? How did they gain control? How did the colonizers change the countries they colonized? When and how did the colonies gain their independence?

- Examine terrorism in different parts of the world. What terrorist groups are in operation? What are their geopolitical objectives? Do they have any geographic claims to territory? Are the groups justified in their political claims? What tactics other than terrorism might be used?

- Select a city in a different part of the world. Describe the historical founding and growth of the city, as well as the current form and structure of the city. How is the city interconnected with the global economy? How has globalization shaped the form and structure of the city?

Practice Quizzes

Answers appear at the end of this book

Vocabulary Matching: Match the term to its definition.

A. Colonialism
B. Core-periphery model
C. Cultural hybridization (or syncretism)
D. Cultural imperialism
E. Cultural nationalism
F. Culture
G. Demographic Transition
H. Economic Growth Rate

I. Ethnic religion
J. Globalization
K. Nation-state
L. Population Pyramid
M. Pull force
N. Purchasing Power Parity
O. Push force
P. Universalizing religion

1. _____ Graphic representation of a country's population structure showing age and gender.

2. _____ Learned and shared behavior by a group of people empowering them with a distinct "way of life"; this includes both material (technology, tools, etc.) and abstract (speech, values, etc.) components.

3. _____ The blending of popular culture with local cultural traditions

4. _____ The growing interconnectedness of people and places through converging processes of economic, political, and cultural change

5. _____ A favorable event or condition in a potential destination that encourages a person to immigrate

6. _____ The four-stage model of population change derived from the historical decline of the natural rate of increase as population becomes increasingly urbanized through industrial and economic development

7. _____ A negative event or condition in a person's homeland that encourages a person to emigrate

8. _____ The active promotion of one cultural system at the expense of another

9. _____ The process of defending a cultural system against offensive cultural expressions, while promoting local cultural values

10. _____ A religion that attempts to appeal to all peoples regardless of location or culture (ex.: Christianity, Islam, Buddhism)

11. _____ A model that holds that the U.S., Canada, Western Europe, and Japan constitute the global economic center, while most of the areas to the south make up a less-developed zone

12. _____ A relatively homogeneous cultural group with its own fully independent political territory

13. _____ The formal establishment of rule over a foreign population

14. _____ A religion closely identified with a specific national or tribal group, often to the point of assuming the role of the major defining characteristic of that group; normally these religions do not actively seek new converts

15. _____ The annual rate of expansion for Gross Domestic Product (GDP)

Multiple Choice Quiz: *Choose the word or phrase that best answers the question*

1. What is the major component of globalization?
   a. Economic reorganization of the world
   b. Global transportation
   c. International athletic competition
   d. Global warming
   e. World communication

2. Which of the following is an example of the globalization of non-material culture?
   a. Eating Thai food in London, England
   b. A Turkish shopkeeper speaking English to an American tourist in Istanbul
   c. A Honda automobile assembly plant in Greensburg, Indiana
   d. A McDonald's in Moscow, Russia
   e. All of the above

3. All of the following statements about globalization are true, EXCEPT...
   a. Cultural globalization is a one-way flow that spreads from the United States to the rest of the world
   b. Globalization has profound geopolitical implications
   c. Globalization is aggravating worldwide environmental problems
   d. Resources previously used only by small local groups are now viewed as global commodities
   e. There is a significant criminal element to contemporary globalization

4. Which of the following people would be LEAST likely to support globalization?
   a. A CEO of a major corporation
   b. A leader of the Republican party
   c. A leader of the Democratic party
   d. A member of a labor union
   e. An economist

5. All of the G8 countries, the exclusive club of the world's major industrial nations, are located in a common part of the world. Which one?
   a. In the tropics
   b. Western Hemisphere
   c. Eastern Hemisphere
   d. Southern Hemisphere
   e. Northern Hemisphere

6. Where are most wars fought today?
   a. Between countries
   b. Globally
   c. In Latin America
   d. In the Middle East
   e. Within countries

7. With what other discipline can geography be readily compared?
   a. Sociology
   b. Political Science
   c. History
   d. Economics
   e. Anthropology

8. According to your textbook, _____ is fundamental to cultural cohesiveness.
   a. Economic well-being
   b. Language
   c. Politics
   d. Religion
   e. Sustainability

9. Which one of the following states would be most likely to be included in the vernacular region of the "Midwest" in the United States?
   a. Arizona
   b. Georgia
   c. Indiana
   d. Maine
   e. Oregon

10. The vast majority of current population growth occurs in four world regions. Which of the following is NOT one of these world regions?
    a. South Asia
    b. Latin America
    c. East Asia
    d. Europe
    e. Africa

11. Which measure of population is produced by subtracting the number of deaths in a given year from the number of births in that same year?
    a. Total Fertility Rate (TFR)
    b. Percent of population under age 15
    c. Rate of Natural Increase (RNI)
    d. Percent of Population over age 65
    e. Migration

12. What does a population pyramid with a wide base and a narrow top tell us about the growth rate of the population represented by the pyramid?
    a. The population is experiencing rapid growth
    b. The population is experiencing slow growth
    c. The population is experiencing no growth
    d. The population is experiencing negative growth
    e. Population pyramids cannot tell us anything about the growth rate

13.　Which of the following countries is NOT one of the top destinations for international migrants?
    a.　United States
    b.　France
    c.　Germany
    d.　Canada
    e.　China

14.　When the modern country of Turkey was established in the early 1900s, its leader (Ataturk) created the Turkish Linguistic Society and charged it with purging the language of foreign phrases. This is an example of which of the following practices?
    a.　Cultural assimilation
    b.　Cultural hybridization
    c.　Cultural imperialism
    d.　Cultural nationalism
    e.　Cultural syncretism

15.　A high birthrate usually goes along with which of the following statistics?
    a.　High female illiteracy
    b.　High GNP per capita
    c.　High life expectancy at birth
    d.　Low under-age-5 mortality
    e.　All of the above

Summary of the 10 Largest Countries

Total Population of the World: About 7 billion

Population Indicators for 10 Largest Countries

	Highest (country and value)	Lowest (country and value)
Population 2010 (millions)	China: 1,318.1	Japan 127.4
Density per sq km	Bangladesh: 1,142	Russia: 8
RNI	Pakistan: 2.3	Russia: -0.2
TFR	Nigeria: 5.7	Japan: 1.4
Percent Urban	Japan: 86%	Bangladesh: 25%
Percent < 15	Nigeria: 43%	Japan: 13%
Percent > 65	Nigeria: 3%	Japan: 23%
Net Migration (per 1000; 2000-05)	United States: 3.3	Pakistan: -1.6

<u>Development Indicators for 10 Largest Countries</u>

	Highest (country and value)	Lowest (country and value)
GNI per capita/PPP 2008	United States: $48,430	Bangladesh: $1,450
GDP Avg. Annual Growth (2000-2008)	China: 10.4%	Japan: 1.1%
Human Development Index (2007)	Japan: 0.960	Nigeria: 0.511
Percent Living below $2/day	Nigeria: 83.9%	U.S., Japan: 0%
Life Expectancy 2010	Japan: 83	Nigeria: 47
> 5 Mortality 2008	Nigeria: 186 per 1000	Japan: 4 per 1000
Gender Equity 2008	Bangladesh: 106	Pakistan: 80

Chapter 2

The Changing Global Environment

Learning Objectives

➢ This chapter provides an overview of Earth's environmental systems, including its geology, climate, hydrology, and biogeography.
➢ The objective is to help you develop a broad understanding of the Earth's environmental systems provides a physical context for understanding regional human activities in subsequent chapters
➢ The chapter identifies and discusses key physical features and their formation.
➢ It also describes basic physical processes, such as volcanic and seismic activities
➢ The chapter introduces and discusses the interrelationships between physical processes and human activities, such as irrigation, resource use and depletion, and global warming
➢ When you complete this chapter, you should have a basic foundation of how Earth's physical systems are crucial in understanding the differences of human activity around the world

Chapter Outline

1. **Geology and Human Settlement**
   A. Plate tectonics: a geophysical theory that postulates that Earth is composed of a large number of geologic plates that move slowly across its surface
      a. **Earth's interior:** separated into three major zones: core, mantle, and outer crust
      b. **Convection cells:** large areas of very slow-moving molten rock within Earth
      c. **Tectonic plates:** huge, continent-sized blocks of rock; similar to jigsaw puzzle pieces
      d. **Convergent plate boundary:** a point where two plates are being forced together by convection cells deep inside Earth
      e. **Subduction zone:** a region where one tectonic plate is pushed and pulled beneath another plate; characterized by deep trenches
      f. **Divergent plate boundary:** a point on Earth where two plates move away from each other in opposite directions; rift valleys, such as the Red Sea, are examples of divergent plate boundaries that form deep depressions on Earth's surface
      g. *Pangaea*: a supercontinent that existed 250 million years ago, centered on present-day Africa
2. **Geologic Hazards: Earthquakes and Volcanoes**
   A. Earthquakes and volcanoes have a major impact on human settlement
   B. Societies most vulnerable to geologic hazards are those where urbanization has taken place rapidly
3. **Global Climates**
   A. **Human–Climate Links**
      a. Human settlement and food production are closely linked to weather and climate
      b. Human activities are changing Earth's climate
   B. **Climatic Controls:** five main factors explain differences in global weather and climate
      a. **Solar energy:** incoming solar energy (*insolation*) that heats the earth
         i. **Greenhouse effect**: a term that describes the natural process of atmosphere heating; insolation is absorbed by land and water surfaces and heats Earth's atmosphere

    b. **Latitude:** insolation is most intense in the equatorial region and becomes less intense toward the poles; this is because the curvature of Earth results in the incoming solar radiation being 'spread' over wider areas

    c. **Interaction between land and water:** land surfaces heat up and cool down more rapidly than water surfaces; because of this, bodies of water are said to moderate climates; oceanic or maritime climates are more moderate and do not exhibit the extreme temperature fluctuations experienced in the interior of continents. The term *continentality* is used to refer to the different heating and cooling properties of land and water.

    d. **Global Pressure Systems:** Atmospheric pressure is uneven across the Earth's surface; this results from the uneven heating of the Earth because of latitudinal differences and the global land--water configuration; differences in atmosphere high- and low-pressure systems contribute to different weather patterns (including storms) across the Earth's surface

    e. **Global wind patterns:** pressure systems produce wind because air (wind) flows from high-pressure areas to low-pressure areas

C. World Climate Regions

    a. **Weather:** short-term, day-to-day (or hour-to-hour) expression of atmospheric conditions, including temperature, pressure, precipitation, humidity, and wind

    b. **Climate:** a long-term (usually at least 30 year) description of an area's weather; a statistical averaging of data on temperature, pressure, precipitation, etc.

    c. **Climate region:** an area within which similar climatic conditions prevail; Figure 2.13 illustrates six broad climate regions, with associated sub-climatic regions

    d. **Climograph:** a visual representation of average high and low temperatures and precipitation amounts at a specific location over the course of a year

D. Global Warming

    a. Human activities, including industrialization, are changing the Earth's climate; human-caused changes are termed *anthropogenic*

    b. **Global warming:** a concept that refers to the overall increase in the temperature of the Earth's atmosphere

        i. Global warming may result in changing weather patterns, sea-level rise, and extinctions to some plants and animals

        ii. As a result of changing weather patterns, the world's food production will also be affected; this is because agricultural practices are closely interconnected with weather and climate

    c. **Causes of global warming:** global warming results from an intensification of the greenhouse effect—the trapping of more incoming and outgoing solar radiation by gases in the atmosphere; four major greenhouse gasses account for global warming

        i. **Carbon dioxide:** an increase in carbon dioxide has resulted primarily from the burning of fossil fuels such as coal and petroleum

        ii. **Chlorofluorocarbons (CFCs):** an increase in CFC emissions has resulted from the widespread use of aerosol sprays and refrigeration

        iii. **Methane:** An increase in methane has resulted from vegetation burning associated with forest clearing, anaerobic activity in flooded fields, cattle and sheep effluent, and leakage from pipelines and refineries connected with natural gas production

        iv. **Nitrous oxide:** an increase in nitrous oxide has resulted from the increased use of industrial fertilizers

    d. **Effects of global warming:** global warming may result in an increase of the Earth's atmosphere by 2 to 4°F by 2030; climate change may cause a shift in agricultural practices, thus endangering food production; increased global temperatures will cause sea levels to rise, thus flooding low-lying coastal locations

    E. Globalization and Global Warming: Global warming has become a geopolitical issue
      a. International restrictions on Greenhouse Gas (GHG) emissions
        i. **1992 Earth Summit in Rio de Janeiro:** 167 countries signed an agreement to voluntarily limit greenhouse gas emissions
        ii. **1997 Kyoto Protocol:** 30 industrialized countries agreed to reduce their emissions; became international law in 2005; the United States never ratified the protocol based on concerns that emission regulation would harm the U.S. economy
        iii. **2009 Copenhagen Accord:** working documents signed by 188 countries; major points include agreement by signatories that global warming is an urgent problem; that action is urgently needed for adapting to effects of global warming; recognizes the importance of local, regional, and national policies for emission reduction; and that emphasis is placed on deforestation and forest degradation
      b. Political obstacles to addressing global warming
        i. Neither China nor India, both of which are significant contributors to greenhouse gas emissions, believe that they must adhere to international emissions reduction plans for fear of harming their economic growth
        ii. The financial aid contribution of more-developed countries to less-developed countries as aid for adaptation to global warming is highly controversial
        iii. It remains unclear if the United Nations is the appropriate agency to forge a global warming agreement

4. Water: A Scarce Global Resource
    A. Water is unevenly distributed around the world
    B. **The Global Water Budget**
      a. More than 70 percent of the Earth's surface is covered by water
      b. 97 percent of the total global water supply is saltwater; only 3 percent is freshwater
      c. Of Earth's freshwater, almost 70 percent is stored in polar ice caps and glaciers; 30 percent occurs as groundwater; less than one percent of the Earth's freshwater is accessible from surface rivers and lakes
      d. **Water stress:** a term used to indicate where water problems exist or might occur; stress relates to situations of water scarcity, sanitation, access, and management
        i. **Water scarcity:** about half the world's population lives in areas where water shortages are common
        ii. **Water sanitation:** polluted water contributes to sickness and death
        iii. **Water access:** not all people have access to clean, fresh water; numerous hardships, borne especially by women, result
        iv. **Water resource management:** the regulation and allocation of water resources; this is a highly political process

5. Human Impacts on Plants and Animals: The Globalization of Nature
    A. **Biomes:** the biogeographic term used to describe a grouping of the world's flora and fauna into a large ecological region; used interchangeably with bioregion
      a. **Tropical Forests and Savannas**
        i. Occur in equatorial climate zones with high average annual temperatures and high amounts of rainfall
        ii. Covers about seven percent of Earth's land surface
        iii. Characterized by three-tiered canopy of vegetation
        iv. Most nutrients are stored in living plants, not the soil
        v. Tropical forests are not well suited for intensive agriculture
      b. **Deforestation in the Tropics**
        i. Tropical forests are being rapidly devastated

        ii.   Rates of deforestation differ from region to region; deforestation is occurring most rapidly in Southeast Asia

       iii.   Deforestation is a political, economical, and ethical problem

       iv.   Tropical deforestation contributes to global warming

       v.   The globalization of commerce in wood products has contributed to widespread deforestation

       vi.   The causes of deforestation include the globalization of the timber industry; the replacement of forests with pastures for cattle; the expansion of human settlements

  c. **Deserts and Grasslands**

       i.   One-third of the Earth's surface is desert, with annual rainfall of less than 10 inches

       ii.   Grasslands include *prairies*, which are characterized by thick, long grasses; and *steppes*, which are characterized by shorter, less dense grasslands. Prairies predominate throughout the midsection of North America, while steppes are commonly found in Central and Southwest Asia

       iii.   Desertification: this refers to the spread of desert-like conditions into semiarid grasslands; desertification has affected many parts of the world, and is caused by overgrazing, poor cropping practices, and the buildup of salts in soils from irrigation; desertification has contributed to an increasing number of severe sandstorms, especially in China

  d. **Temperate Forests**

       i.   Temperate forests are large tracts of forests found in middle and high latitudes

       ii.   Two types of trees dominate: softwood coniferous, or evergreen, trees (e.g., pine, spruce, and fir); and hardwood deciduous trees (e.g., elm, maple, and beech); hardwood trees are more difficult to mill and are thus less favored by the timber industry

       iii.   The struggle between the interests of the timber industry and of environmental groups is very political

       iv.   Globalizing processes, including the worldwide demand for and consumption of timber products, play a significant role in the transformation of local ecosystems

6. **Food Resources: Environment, Diversity, and Globalization**

  A. **Population increases and land pressures:** every minute, approximately 258 people are born in the world; but every minute, about 10 acres of cropland are lost because of environmental problems such as soil erosion and desertification

  B. **Green Revolutions:** a term that refers to innovations in agricultural techniques, often using genetically altered seeds coupled with high inputs of chemical fertilizers and pesticides

    a. There have been two stages of green revolutions. The first stage (1950–1970) combined three processes: a change from traditional mixed crops to monocrops; intensive applications of irrigated water, fertilizers, and pesticides; and additional increases in the intensity of agriculture by reducing the fallow, or field-resting time, between seasonal crops. The second stage (since the 1970s) emphasizes new strains of fast-growing wheat and rice for tropical and subtropical climates; these new strains allow farmers to grow two or three crops per year on lands that previously supported only one crop per year.

    b. **Environmental and social costs:** green revolutions have led to an increase of fossil fuels and thus contributes to global warming; widespread habitat and wildlife loss has been associated with green revolutions; local ecosystems have suffered increased pollution; green revolutions may not be economically sustainable; economic disparities between those who benefit from green revolution technology and those who do not have led to political tensions

C. **Problems and Projections:** four general points are raised with respect to the world's food security

    a. Local issues, such as poverty, civil unrest, and war, often deny people access to their ability to grow, buy, or receive adequate food

    b. Political problems are more commonly responsible for food shortages and famines than are natural events such as drought and flooding; food distribution, both locally and globally, is highly politicized

    c. Globalization is causing dietary preferences to change: an increasing number of people world-wide are including more meat in their diets, a shift that is both economic-based and culturally-influenced; these dietary shifts may not be sustainable for the world's population

    d. Africa and South Asia are areas of greatest concern regarding food; problems stem from rapid population growth, civil disruption, and warfare

Summary

- Global environmental change is driven by human activities
- Globalization is both a help and a hindrance to world environmental problems
- The arrangement of tectonic plates on Earth is responsible both for diverse physical landscapes and also for geologic hazards that threaten the well-being of human settlements
- Climate change and global warming is a by-product of industrialization
- Plants and animals throughout the world face an extinction crisis because of habitat destruction resultant from human activities; tropical forest ecosystems are especially vulnerable
- Although the world's population is growing more slowly than in previous decades, and even though the world's food supplies have increased, accessibility to adequate food remains a problem for much of the world's population

Research or Term Paper Ideas

- On December 26, 2004 a major earthquake occurred along the Indian and Burma plate boundary underneath the Indian Ocean. This seismic event resulted in a catastrophic tsunami that caused the death of hundreds of thousands of people in places from Southeast Asia to the eastern coast of Africa. Contrast how the tsunami affected different places throughout the region, and how different governments have been able to rebuild following this disaster.

- Tropical deforestation is a major environmental problem. Select a country, such as Brazil or Cambodia, and learn how various practices of globalization are contributing to the deforestation. How are these practices interconnected?

- One consequence of global warming is a rise in sea level. What would happen to coastal cities in North America, such as Miami and New Orleans, following such a rise in sea level?

Practice Quizzes

Answers appear at the end of this book.

Vocabulary Matching: *Match the term to its definition.*

A. Anthropogenic
B. Bioregion
C. Climate region
D. Climograph
E. Continentality
F. Convergent plate boundaries
G. Desertification
H. Divergent plate boundaries

H. Global warming
I. Green revolution
J. Insolation
K. Maritime climate
L. Plate tectonics
M. Prairie
N. Steppe
O. Water stress

1. _____ Caused by humans.

2. _____ The theory that explains the gradual movement of large geographical platforms along Earth's surface; this theory postulates that Earth is made up of a large number of geological plates that move slowly across its surface; it helps explain both the inner workings and the surface landscape features of Earth.

3. _____ A place where two tectonic plates are forced together by convection cells deep within Earth, often resulting in collision of plates that create mountains; the Cascade Range in Washington and Oregon is an example.

4. _____ A place where two tectonic plates move in opposite directions, which often allows magma to flow to Earth's surface.

5. _____ An assemblage of local plants, animals, and insects covering a large area such as a tropical rainforest or a grassland.

6. _____ Spread of desert-like conditions into semi-arid areas.

7. _____ Agricultural program that began in the 1950s and involved new agricultural techniques using genetically altered seeds, along with high inputs of chemical fertilizers and pesticides.

8. _____ An area on a map where similar climatic conditions prevail over a larger area.

9. _____ Describes climate regions in the interior of continents, where they are removed from moderating ocean influences, and characterized by hot summers and cold winters; at least one month must average below freezing.

10. _____ Climate moderated by proximity to oceans or large seas; it is usually cool, cloudy, and wet, without temperature extremes.

11. _____ Graph of average high and low temperatures and precipitation for an entire year (usually provided for a specific city).

12. _____ Incoming solar energy that enters the atmosphere adjacent to Earth.

13. _____ An increase in the average temperature of Earth's atmosphere.

14. _____ An environmental planning tool used to predict areas that have – or will have – serious water problems based on the per capita demand and supply of freshwater.

15. _____ A vegetative region characterized by shorter, less dense grasslands.

Multiple Choice: *Choose the word or phrase that best answers the question.*

1. Why is the theory of plate tectonics important?
   a. It explains and describes the inner workings of Earth
   b. It explains and describes many of Earth's surface landscape features
   c. It gives clues about the world distribution of earthquakes and volcanoes
   d. A and B above
   e. A, B, and C above

2. Where does the heat exchange described in plate tectonic theory take place?
   a. At the convergent plate boundary
   b. At the divergent plate boundary
   c. In convection cells
   d. In the subduction zone
   e. In the center of the tectonic plates

3. Which of the following natural hazards is hardest to predict?
   a. Earthquakes
   b. Hurricanes
   c. Flooding
   d. Blizzards
   e. Volcanic eruptions

4. Which of the following statements about volcanoes is FALSE?
   a. The loss of life from volcanoes is a fraction of that from earthquakes
   b. Volcanoes are found along most tectonic plate boundaries
   c. Volcanoes can cause major destruction
   d. Volcanic eruptions cannot be predicted
   e. Volcanoes provide some benefits

5. What is the major factor that works to control Earth's climate?
   a. Latitude
   b. Insolation
   c. Interaction between land and water
   d. Global pressure systems and wind patterns
   e. All of the above

6. Which of the following cities is most subject to continentality in its climate?
   a. Duluth, Minnesota
   b. Miami, Florida
   c. San Diego, California
   d. Seattle, Washington
   e. Virginia Beach, Virginia

7.   What does the Köppen system describe?
       a. Climate
       b. Elevation
       c. Landforms
       d. Population
       e. Soils

8.   Which of the following statements about the Greenhouse Effect is FALSE?
       a. Emissions in the lower atmosphere are increasing the greenhouse effect
       b. Carbon dioxide accounts for more than half of the human-generated greenhouse gasses
       c. Natural greenhouse gasses have varied over long periods of geologic time
       d. The composition of greenhouse gasses has changed dramatically because of industrialization
       e. The Earth would be far better off if we could eliminate the greenhouse effect

9.   Which of these accounts for more than half of the human-generated greenhouse gases?
       a. Ozone
       b. Nitrous Oxide
       c. Methane
       d. Chlorofluorocarbons
       e. Carbon dioxide

10.   How much of the water on Earth is freshwater?
       a. 3%
       b. 8%
       c. 14%
       d. 18%
       e. 22%

11.   What is the greatest source of illness and death worldwide?
       a. Contagious diseases
       b. Avian flu
       c. Cancer
       d. Heart disease
       e. Polluted water sources

12.   This activity is believed to be a major cause of desertification.
       a. Clear-cutting forests
       b. Farming marginal lands
       c. Implementing Green Revolution strategies
       d. The release of greenhouse gases
       e. All of the above

13.   How much of the world's land qualifies as true desert?
       a. One-tenth
       b. One-eighth
       c. One-fourth
       d. One-third
       f. One-half

14.   Traditional intensive farming generally requires large inputs of this resource.
    a. Fertilizer
    b. Herbicides
    c. Labor
    d. Pesticides
    e. Irrigation

15.   Which of the following is NOT an environmental cost of the Green Revolution?
    a. Stratification of society based on wealth and poverty
    b. Pollution of rivers and water sources
    c. Increased air pollution from factories producing agricultural chemicals
    d. Damage to wildlife
    e. Destruction of habitat

Chapter 3

North America

Learning Objectives

➢ This chapter introduces the United States and Canada, the two countries comprising North America
➢ The impact of globalization, both as it impacts and is impacted by North America is emphasized
➢ The widespread abundance and affluence of North America is contrasted with persisting disparities in income and quality
➢ North America's unique cultural character and heritage is detailed, stressing its colonial and immigrant heritage
➢ Upon completing this chapter, you should be familiar with the environmental, cultural, political and economic similarities and differences of North America
➢ Key concepts include urban structural models, urban processes, cultural identities and assimilation

Chapter Outline

1. **Introduction:** North America is traditionally defined as including the United States and Canada
   A. North America has been significantly affected by globalization, but the region has also contributed to processes of globalization
   B. North America is a culturally diverse and resource-rich region
      a. North America is highly urbanized and has one of the world's highest rates of resource consumption
      b. The region exemplifies a post-industrial economy
   C. The term "North America" is problematic; the continent of North America includes both Mexico and Central America, and often the Caribbean; culturally, however, the political border between the United States and Mexico has been used to differentiate the region
   D. North America reflects significant economic disparities
   E. North America has a rich ethnic and cultural heritage
2. **Environmental Geography: A Threatened Land of Plenty**
   A. North America's physical and human geographies are diverse and intricately linked
      a. **The Costs of Human Modification**
         i. North America's physical setting has been highly modified
         ii. Globalization and urban/economic growth have transformed North America's landforms, soils, vegetation, and climate
      b. **Transforming Soils and Vegetation**
         i. The arrival of Europeans impacted the region's flora and fauna; new species, such as wheat, cattle, and horses, were introduced; forest cover was removed and grasslands were converted to farm land
      c. **Managing Water**
         i. Water consumption is high in North America
         ii. Many aquifers, such as the Ogalla Aquifer, are being depleted

          iii. Water quality is a major issue; despite the introduction of environmental laws and guidelines, many North Americans are exposed to various water-borne pollutants

          iv. Disasters such as the oil spill resultant from the explosion on the Deepwater Horizon oil rig continue to pose environmental hazards

          v. North America's fisheries are threatened from overfishing, environmental degradation, infections, and the introduction of exotic species

  d. **Altering the Atmosphere**

          i. North Americans have modified both local and regional climates; the chemical composition of Earth's atmosphere has also been affected by urban and industrial growth in North America

          ii. North America's urban and industrial growth has contributed to transnational environmental problems, such as acid rain

  e. **The Price of Affluence**: Globalization has provided many benefits, but not without significant costs

          i. Energy consumption remains high, leading to environmental and economic costs

          ii. Environmental initiatives have addressed local and regional problems

  f. There is increasing support to address these problems

          i. Sustainable agriculture, whereby organic farming principles, limited use of chemicals, and an integrated plan of crop and livestock management, is increasing in popularity

          ii. The development of renewable energy sources, such as hydroelectric, solar, wind, and geothermal power has been encouraged; fossil fuels, however, continue to dominate energy consumption

B. **A Diverse Physical Setting**

  1. North America's landscape is dominated by interior lowlands and bordered by mountainous topography in the West; coastal plains dominate the eastern portion

  2. The Atlantic coastline is composed of drowned river valleys, bays, swamps, and barrier islands

  3. The Piedmont, Appalachian Highlands, and interior highlands (e.g., Ozark Mountains and Ouachita Plateau) dominate the eastern interior

  4. The interior of North America is dominated by the Great Plains; glacial forces shaped much of this region

  5. Western North America's topography is significantly different from eastern North America; this region is characterized by mountain-building processes, alpine glaciations and erosion processes; key features include the Rocky Mountains, Alaska's Brooks Range, the Colorado Plateau, and Nevada's basin and range landscape

  6. North America's western border includes the Coast Ranges of Washington, Oregon, and California; these are interspersed with the lowlands of the Puget Sound (Washington), Willamette Valley (Oregon), and Central Valley (California)

C. **Patterns of Climate and Vegetation**

  1. North America's climates and vegetation are diverse, resultant from the region's size, latitudinal range, and topography

  2. Environments south of the Great Lakes are characterized by longer growing seasons and deciduous broadleaf forests; environments north of the Great

Lakes are dominated by coniferous evergreen, or boreal, forests; the far north is characterized by tundra—a mixture of low shrubs, grasses, and flowering herbs

3. Much of the interior of North America is dominated by drier continental climates; the soils of this region are fertile and originally supported prairie vegetation

4. A marine west coast climate dominates the western coastal region north of San Francisco; the southwest coastal regions are dominated by a Mediterranean climate

D. **Global Warming in North America**
1. Global warming has already impacted North America
   a. High latitude and alpine environments have been modified because of global warming, impacting whale and polar bear populations; traditional ways of life among North America's native populations, such as the Inuit, have been affected
   b. Changes in sea-level have increased coastal erosion
2. The long-term consequences of climate change on North America are enormous
   a. Many coast locations will be more vulnerable to rising sea levels; distributions of flora and fauna will be changed; many alpine glaciers are disappearing; agricultural patterns and practices will be transformed; weather patterns will be modified
   b. In the United States, federal policies enacted in response to global climate change will impact economic and business practices
      i. A cap-and-trade energy policy, in which overall carbon emissions would be limited or capped within a system that would allow polluters to buy and sell carbon credits in an emissions-trading marketplace, might reduce overall levels of emissions

3. **Population and Settlement**
A. **Modern Spatial and Demographic Patterns**
1. Metropolitan clusters dominate North America's population geography; settlement is unevenly distributed across the region
2. In Canada, 90 percent of the country's population is located within 100 miles of the U.S. border; this is known as Canada's "Main Street" corridor and includes the country's major cities of Toronto, Montreal, Ottawa, and Hamilton
3. The largest agglomeration of people in the United States reside in "Megalopolis," an urbanized area that includes Boston, Washington, D.C., Philadelphia, and New York City; other major urban areas are located around the Great Lakes (including Chicago, Detroit, Cleveland, and Buffalo) and the West Coast (including San Diego, Los Angeles, San Francisco, and Seattle)
4. North America's population has increased; prior to the 20th century, population growth was fueled by high birth rates and immigration; current population growth is the result primarily of immigration

B. **Occupying the Land**
1. North America was occupied by indigenous peoples (known as Native Americans in the United States and the First Nations in Canada) for at least 12,000 years prior to the arrival of Europeans
2. Indigenous peoples were broadly distributed throughout the region; their cultures varied tremendously and reflected adaptations to the different environments of North America; there were an estimated 3.2 million Native

Americans and 1.2 million First Nations people prior to the arrival of Europeans; the introduction of diseases and other disruptions decimated these populations

3. There are three stages of European settlement: Stage I (c. 1600–1750) is marked by the establishment of European colonies, mostly along the coastal regions of eastern North America; Stage II (1750–1850) is marked by the expansion and infilling of settlements throughout prime agricultural areas of eastern North America; Stage III (1850–1910) is marked by an acceleration and westward expansion of settlement as new lands were sought for both settlement and resource use; this was one of the world's largest and most rapid transformations of landscape change

C. **North American Migration**

1. **Westward-Moving Populations:** the most persistent migration trend in North America has been a westward movement; many of the most rapidly growing states are located in the American West (e.g., Arizona and Nevada) and the western Canadian provinces of Alberta and British Columbia; in many western regions, high-technology industries and services, as well as the region's scenic, recreational, and retirement amenities, have contributed to patterns of in-migration; the recent economic slow-down has led to a decline in growth in some western cities, such as Las Vegas and Phoenix

2. **Black Exodus from the South:** prior to the American Civil War, most African Americans lived in the southern United States; since the abolition of slavery and the end of the American Civil War, many African Americans have left the South, moving to urban areas in the North (such as Chicago, Philadelphia, Detroit, and New York City) or to western cities (such as Los Angeles and Oakland); this early exodus was the result of declining economic opportunities in the South, coupled with increased industrial jobs in the North; beginning in the 1970s, there was a slight reversal of this trend, as many African Americans moved from northern to southern locations

3. **Rural-to-Urban Migration**: historically, people in North America have moved from the countryside to the city; shifting economic opportunities account for much of the transformation

4. **Growth of the Sunbelt South:** southern states from the Carolinas to Texas have experienced considerable growth throughout the 21st century; key metropolitan areas, such as Raleigh, North Carolina, and Austin, Texas, have received high numbers of domestic migrants; numerous factors account for this growth, including a more buoyant economy, modest cost-of-living, and recreational opportunities

5. **Nonmetropolitan Growth:** this is a process in which people leave large cities and move to smaller towns and rural areas; this trend has been most noticeable since the 1970s and reflects, in part, the decisions of "lifestyle" migrants; these are people who find or create employment opportunities in more affordable smaller cities and rural settings that are rich in amenities and perceived to be free of the problems (e.g., crime) of larger, urban areas.

D. **Settlement Geographies:** The Decentralized Metropolis

1. North American settlement patterns largely reflect a process of *urban decentralization*, in which metropolitan areas sprawl in all directions and suburbs take on many of the characteristics of traditional downtowns; the impact of urban decentralization is more pronounced in the U.S. than it is in Canada; some observers identify a globalization of urban decentralization, suggesting that many Asian, European, and Latin American cities are undergoing similar transformations

2. **Historical Evolution of the City in the United States:** U.S. cities have been shaped in large part by changing transportation technologies; four technological stages are identified: pedestrian/horse-car technology (pre-1888), resulting in compact cities; electric trolley/street car (1888–1920), resulting in expanded, star-shaped urban patterns; automobile (1920–1945), resulting in sprawling cities; and freeway (1945–present), resulting in decentralized urban areas, with suburbs extended up to 60 miles from the traditional downtown

    a. **Concentric Zone Model:** an idealized urban land use model that presupposes urban areas are spatially organized in a series of rings (or zones); this model accounts for urban form of early 20th century North American cities

    b. **Urban realms model:** an urban model that accounts for the increased suburbanization of North American cities, including the existence of peripheral retailing, industrial parks, office complexes, and entertainment facilities; also called "edge cities," these spatially distant urban places have fewer functional connections with the central city

3. **The Consequences of Sprawl:** suburbanization continues to shape and reshape the evolution of the North American city

    a. Processes of suburbanization have contributed to many problems with the inner city; population losses have contributed to a shrinking tax base; poverty rates and unemployment remain high; and ongoing discrimination, segregation, and poverty contribute to social and racial tensions

    b. **Gentrification:** a process in which higher-income residents move back to the inner city, thereby displacing lower-income residents; coupled with the construction of new shopping complexes, entertainment attractions, or convention centers; inner-city locations are rehabilitated, but this includes a social cost to poorer populations

    c. **New urbanism:** this term refers to an urban design movement that stresses higher density, mixed-use, pedestrian-scaled neighborhoods

    d. **Suburban downtowns:** similar to edge cities, suburbs are becoming full-service urban centers; key corporate offices are increasingly relocating to these locations; these locations are intimately tied to global flows of information, technology, capital, and migrants

E. **Settlement Geographies: Rural North America**

1. Rural settlements in North America originated with European settlement; patterns in the United States reflect the federal government's township-and-range survey system; introduced in 1785, this system divided lands into six-mile-square townships, which were then sold to the public; Canada developed a similar system of surveys; the resultant rural landscape following these systems was of a rectilinear character

2. The advent of the railroad transformed the rural landscape; corridors of development were opened, thereby providing access to markets for commercial crops; new towns were established as a result

3. Throughout the 20th and 21st centuries, family farms have been replaced by corporate farms; consequently, the population of many rural areas has decreased, as people move toward urban areas with greater employment opportunities

4. Some rural areas, particularly those associated with the growth of "edge cities" have experienced growth

4. **Cultural Coherence and Diversity**
    A. North America's cultural geography is globally dominant and internally pluralistic
    B. **The Roots of a Cultural Identity**
        1. Early dominance of British culture; traditional beliefs associated with representative government, separation of church and state, liberal individualism, privacy, pragmatism, and social mobility
        2. Consumer culture flourished after 1920; values center around convenience, consumption, and the mass media
        3. North America's cultural unity coexists with pluralism—the persistence and assertion of distinctive cultural identities; related to the concept of *ethnicity*, a term denoting a group of people with a common background
    C. **Peopling North America**
        1. North America is a region of immigrants; beginning in the 17th century, indigenous peoples have been displaced; varied immigrant groups have produced a culturally diverse landscape throughout North America; the region is largely defined by different degrees of *cultural assimilation*, the process in which immigrants are absorbed into the larger host society
        2. Migration to the United States has occurred in five phrases: Phase I (pre-1820) included mostly English settlers and enslaved Africans; Phase II (1820–1870) included immigrants primarily from northwestern Europe—especially Irish and Germans; Phase III (1870–1920) included immigrants mostly from southern and eastern Europe; Phase IV (1920–1970), in which most immigrants originated from neighboring Canada but especially Latin America; owing to immigration restrictions, overall immigration levels declined; Phase V (1970s–present), in which immigration has intensified, with many immigrants originating from Asia and Latin America
        3. The Hispanic population of the United States has increased rapidly; migrants from Latin America continue to move to the southern border states of California, Arizona, New Mexico, and Texas, but also in other states such as Georgia, Kansas, and Arkansas
        4. In percentage terms, migrants from Asia constitute the fastest-growing immigrant group; California remains a key destination for migrants from Asia, especially in the larger cities of San Francisco and Los Angeles; other key communities include New York City, Washington, D.C., Seattle, Chicago, and Houston
        5. In Canada, early French arrivals concentrated in the St. Lawrence Valley; after 1765, European immigrant patterns into Canada were similar to those of the United States; currently, most immigrants to Canada originate from Asia; cities along Canada's west coast, such as Vancouver, are prime recipients of Asian immigrants
    D. **Culture and Place in North America**
        1. **Cultural homeland:** this is a culturally distinctive nucleus of settlement in a well-defined geographic area; it is a cultural landscape with an enduring personality; there are many cultural homelands in North America, including French Canadian Quebec, the Hispanic Borderlands of the American southwest, and the Black Belt of the American south
        2. **A Mosaic of Ethnic Neighborhoods:** North American settlement patterns reveal numerous smaller-scaled and often closely knit ethnic neighborhoods; these ethnic neighborhoods are the result of various social and economic processes

E. **Patterns of North American Religion**
1. The United States exhibits tremendous religious diversity; although Protestantism is the dominant religion, the cultural landscape is also composed of significant concentrations of Catholics, Orthodox Christians, Jews, Buddhists, and Hindus; these different religious groups are often concentrated in specific locales—Orthodox Christians, for example, are especially pronounced in the Northeast
2. Canada's religious diversity is similar to that of the United States; Catholicism in Canada is highly concentrated in French Canadian Quebec

F. **The Globalization of American Culture:** North America is becoming more globalized, while other global cultures are becoming more North American; new hybrid cultures are being created
1. **North Americans: Living Globally:** many foreigners arrive in North America, as migrants and as students; North Americans are exposed to global cultures also through the Internet, international travel, and the consumption of global products; globalization in North America is also creating challenges, such as concerns over language; in some places, hybrid languages, such as *Spanglish*, have emerged
2. **The Global Diffusion of U.S. Culture:** after the Second World War, especially with the Marshall Plan and the Peace Corps, U.S. culture has diffused globally; transportation and information technologies have also facilitated the diffusion of American culture; corporate (e.g., Walt Disney) and media activities (e.g., Time Warner, CNN, and MTV) have facilitated the diffusion of U.S. culture; religious missionaries from the U.S. have also helped diffuse American beliefs and attitudes; many places have resisted or challenged the diffusion of U.S. cultural influences—Canada, for example, has responded to the dominance of U.S. influences while Iran has banned many U.S. films

5. **Geopolitical Framework: Patterns of Dominance and Division**
A. **Creating Political Space:** The U.S. and Canada have different origins
1. **Uniting the States:** A Native-American controlled political space was overwritten by a European political space; 13 English colonies emerged and later achieved independence following a revolution against the English, and then embarked on a westward expansion to its present configuration
2. **Assembling the Provinces:** the Canadian Confederation grew in a piecemeal fashion as provinces were assembled in a slow and uncertain fashion; unlike the U.S., Canada's spatial growth emerged more out of geographical convenience than a compelling nationalism

B. **Continental Neighbors:** geopolitical relationships between the U.S. and Canada have always been close and impact both economic and environmental issues; they share a 5,525 mile border
1. The United States and Canada participate in the International Joint Commission (created out of the Boundary Waters Treaty of 1909); this Commission regulates cross-boundary issues involving Great Lakes water resources, transportation, and environmental quality; other international agreements include the Great Lakes Water Quality Agreement (1972) and the U.S.–Canada Air Quality Agreement (1991).
2. The U.S. and Canada are key trading partners; with Mexico, these three countries signed the North American Free Trade Agreement (NAFTA) in 1994

        3.   Political conflicts still exist: environmental issues are especially important; immigration restrictions have intensified; agricultural and natural resource competition also causes tension

C. **The Legacy of Federalism**
        1.   Two types of state (country) government: (1) *federal*, in which considerable political power and autonomy are allocated to sub-national units of government; and (2) *unitary*, in which power is centralized at the national level; both the United States and Canada exhibit federal forms of political authority, although their origins and evolution are very different
                a.   The United States Constitution initially limited central authority and allocated all unspecified powers to the states; over time, political power has become increasingly centralized at the national level
                b.   The Canadian Constitution initially reserved most powers to central authorizes; it has evolved to reflect more provincial autonomy and a relatively weak national government
        2.   Quebec remains a major political issue in Canada; there are economic and cultural differences between the province of Quebec and the remainder of Canada; many residents of Quebec seek increased autonomy rather than separatism

D. **Native Peoples and National Politics**
        1.   In the U.S., Native American political power has increased; since 1975 with the passage of the Indian Self-Determination and Education Assistance Act in the U.S., Native Americans have gained political autonomy; the Indian Gaming Regulatory Act (1988) offered potential economic independence
        2.   In Canada, the First Nations peoples have gained political autonomy; the Native Claims Office was established in 1975 and has since turned over millions of acres of land to aboriginal control; Canada also created the eastern territory of Nunavut in 1999, representing a new level of native self-government; efforts are underway to create a similar territory in the west

E. **The Politics of U.S. Immigration:** Immigration policies are highly controversial in the U.S.; four key issues are at the center of the debate
        1.   There are continuing disagreements over the legal limit of immigration; some groups support greater restrictions, thus reducing immigration, while other groups favor fewer restrictions, thereby increasing immigration; both groups maintain that their policies would benefit the economy
        2.   There are disagreements over the policing of the U.S.–Mexico border and the need to reduce the in-migration of undocumented migrants
        3.   Immigration policy has become connected to discussions and concerns over the international drug trade and especially drug-related violence along the border
        4.   There is no political consensus on policies addressing the millions of undocumented migrants already residing in the U.S.; some policy-makers advocate amnesty programs, while others advocate stricter felony-level penalties and deportation

F. **A Global Reach:** America's geopolitical influence has expanded over the last two centuries
        1.   The Monroe Doctrine (1824) asserted that U.S. interests were hemispheric and transcended national boundaries
        2.   The Spanish-American War (1898) contributed to further geopolitical expansion, marked by the annexation of the Philippines, Guam, and the Hawaiian Islands

3. Between 1898 and 1916, America's geopolitical influence extended into Central America and the Caribbean
4. During the 1920s and 1930s, the U.S. adopted a relatively isolationist approach to geopolitics; this changed after the Second World War
5. The U.S. extended its geopolitical reach during the Cold War; specific events include the Truman Doctrine, the establishment of the North Atlantic Treaty Organization (NATO), and the Organization of American States (OAS); the U.S. was involved in many regional conflicts and wars, including Korea and Vietnam
6. Following the Cold War, American geopolitical influence intensified in the Middle East; other locations of military involvement include the former Yugoslavia; the U.S. military has adopted a more flexible approach to planning and basing of troops

6. **Economic and Social Development**
   A. **An Abundant Resource Base:** North America has abundant natural resources; its climate provides a diverse agricultural base
      1. **Opportunities for Agriculture:** North America has one of the most efficient food-producing systems in the world; agriculture remains a dominant land use; agriculture is highly commercialized, mechanized, and specialized; the geography of North American farming represents the combined impacts of (1) diverse environments, (2) varied continental and global markets, (3) historical patterns of settlement and agricultural evolution, and (4) the growth role of *agribusiness*, or corporate farming
      2. **Industrial Raw Materials:** North Americans extract and consume huge quantities of natural resources; although well-endowed with many resources, the diverse needs of North American industries require additional materials to be imported; despite significant oil production, U.S. consumption levels require additional and sizeable amounts to be imported; coal is abundant in the U.S., but its overall importance to energy production has declined; North America is endowed with considerable reserves of copper, lead, and zinc and is an important provider of gold, silver, and nickel
   B. **Creating a Continental Economy:** North America's global reach has extended through its economic innovations
      1. **Connectivity and Economic Growth:** *connectivity*—how well different locations are linked through transportation and communication networks; North America's connectivity has increased dramatically throughout its history, facilitating economic growth; technological innovations contributed to increased connectivity
      2. **The Sectoral Transformation:** this refers to the evolution of a nation's labor from one economic sector to another—economic sectors are classified as *primary* (natural resource extraction), *secondary* (manufacturing or industrial), *tertiary* (services), and *quaternary* (information processing); in North America, most workers are employed in the tertiary and quaternary sectors; this represents a transformation from earlier centuries when primary activities, and then secondary activities, were predominant
      3. **Regional Economic Patterns:** *location factors* are the varied influences that explain why an economic activity is located where it is; factors include proximity to natural resources, connectivity, labor supplies, market demand, and capital investment; major manufacturing regions include Megalopolis, Sun Belt locations, the West Coast; many industries benefit from *agglomeration economies*, in which companies with similar and often integrated manufacturing operations locate near one another

C. **Persisting Social Issues:** economic and social problems continue to shape North America's human geography; significant income differences persist and have actually increased; disparities in access to health care and education exist; problems with gender inequity and aging populations remain; globalizing processes have impacted these conditions

1. **Wealth and Poverty:** disparities between the rich and the poor are visible on the landscape; the distribution of wealth and poverty varies widely across the region; many of the wealthiest communities are located in suburban locations, while the poorest are located in the inner-cities

2. **Additional 21ˢᵗ Century Challenges:** companies and workers are attempting to adjust to the uncertainties of a global economy; education is a major public policy issue; gender remains a key social issue, with a significant *gender gap*—differences between men and women in salary issues, working conditions, and political power—remaining; health care and aging are key concerns; the problems of obesity are increasingly recognized as key problems

D. **North America and the Global Economy:** Coupled with Europe, North America plays a pivotal role in the global economy; the region is home to a number of "global cities"; U.S. and Canadian governments and firms played a formative role in creating much of the global economy, as seen in the establishment of the International Monetary Fund, the World Bank, and the General Agreement on Tariffs and Trade (later renamed the World Trade Organization); both the U.S. and Canada also participate in the *Group of Eight* (G-8), a collection of powerful countries that confer regularly on key global economic and political issues.

1. **Patterns of Trade:** both the U.S. and Canada import products from around the world; growth in trade is especially pronounced with Asian countries (e.g., China and South Korea); Canada trades most heavily with the United States; the U.S. exports predominantly automobiles, aircraft, computer and telecommunications equipment, entertainment, financial and tourism services, and food products; Canada mostly exports raw materials (e.g., grain, energy, metals, and wood products) and manufacturing goods

2. **Patterns of Global Investment:** North America attracts huge inflows of foreign capital, both as investment in stocks and bonds, and as foreign direct investment; the U.S. is the largest destination of foreign investment in the world; U.S. citizens invest heavily in Japanese, European, and "emerging" stock markets; three key shifts in broader patterns of globalization are apparent: (1) traditional American-based multinational corporations are adopting new, more globally integrated models; (2) a growing number of multinational corporations are buying companies and assets once controlled by North American or European capital; and (3) many of these companies are also investing heavily in their own, or other regions, of the world

    a. **Outsourcing:** a business practice that transfers portions of a company's production and service activities to lower-cost settings, often overseas; many North Americans have benefitted from cheaper imports as a result of global outsourcing; however, many North Americans have become unemployed because of the transfer of jobs to other countries

Summary

- North Americans have reaped the benefits of the natural abundance of their region; in turn, they have transformed the environment and created a highly affluent society
- North America's affluence has come with a price; the region confronts significant environmental challenges
- In a rather short period of time, a unique cultural imprint has been forged on the region
- North America exhibits two closely intertwined, yet very distinctive, political and cultural trends
- In the United States especially many social problems remain, including ethnic diversity, immigration, poverty, and racial discrimination
- The global economic recession has profoundly impacted the region's economic geography, particularly with respect to housing and unemployment

Research or Term Paper Ideas

- Locate old maps of your city's settlement history—the local library is an excellent source. How has the evolution of your city been influenced by changing transportation patterns? How has the shape and size of your city been impacted? Have these transportation changes been associated with any specific migration, such as the exodus of African Americans from the south?

- Choose a U.S. multinational firm and trace its global activities, including any factors or branch plants it might own in other countries. How extensive are its international operations? What are the benefits and costs of such activities to the firm, to its stakeholders, to U.S. citizens, and to the citizens of the countries in which it is located?

- Study the physical geography of the United States and Canada. What potential alternative energy resources might these two countries develop? Where would these be located?

- Has your city, or perhaps the nearest large city, undergone processes of gentrification? Which parts of the city have been gentrified? Have the benefits of gentrification been experienced broadly, or only to a small group of residents?

Practice Quizzes

Answers appear at the end of this book

Vocabulary Matching: Match the term to its definition.

A. Acid rain
B. Concentric zone model
C. Cultural assimilation
D. Ethnicity
E. Federal states
F. Megalopolis
G. Nonmetropolitan growth
H. Outsourcing

I. Postindustrial economy
J. Renewable energy resources
K. Sectoral transformation
L. Sustainable agriculture
M. Tundra
N. Unitary States
O. Urban decentralization
P. Urban realms model

1. _____ Harmful form of precipitation high in sulfur and nitrogen oxides; it usually originates in industrial regions, but can fall wherever the wind blows it.

2. _____ A mixture of low shrubs, grasses, and flowering herbs that grow in the short growing seasons of the high latitudes.

3. _____ Movement of people from cities to smaller towns and rural areas.

4. _____ Simplified description of urban land use, featuring a number of dispersed, peripheral centers of dynamic commercial and industrial activity linked by sophisticated urban transportation networks.

5. _____ Energy sources that are replenished by nature, including solar, wind, hydropower, and geothermal energy.

6. _____ A large urban region formed as multiple cities grow and merge with one another; the original runs from Boston to Washington, D.C. and includes Baltimore, Philadelphia, and New York City.

7. _____ Environmentally friendly farming that uses organic farming principles, limited use of chemicals, integrated crop and livestock management.

8. _____ Simplified description of urban land use that shows a well-developed central business district surrounded by rings of residences, with higher-income groups living on the urban periphery.

9. _____ Settlement pattern in which metropolitan areas sprawl in all directions and suburbs take on many of the characteristics of traditional downtowns.

10. _____ The evolution of a labor force from being highly dependent on the primary (extraction and agriculture) sector to being oriented around more employment in the secondary, tertiary, and quaternary sectors.

11. _____ The state of an evolving economy wherein traditional manufacturing activity has given way to the growth of high-tech industry and an employment emphasis on services, government, and management information systems.

12. ____   Political system in which a significant amount of power is given to individual states.

13. ____   The process in which immigrants are culturally absorbed into the larger host society.

14. ____   A shared cultural identity held by a group of people with a common background or history, often as a minority group within a larger society.

15. ____   The external purchase of services rather than production in-house; call centers in India are an example.

Multiple Choice: *Choose the word or phrase that best answers the question*

1.   Which of the following statements about North America is/are true?
   a. With a population of about 340 million, North America has the highest rates of resource consumption on Earth.
   b. It is rich in natural resources and is one of the world's most affluent regions.
   c. It is home to 30% of the world's population and consumes 30% of its resources.
   d. A and B above
   e. A, B, and C above

2.   How did European settlers shape the ecology of North America?
   a. They brought new animals, including cattle and horses
   b. They replaced natural grasslands with non-native grain and forage crops
   c. They removed forest cover from millions of acres of land
   d. A and B above
   e. A, B, and C above

3.   Which of the following statements about the Ogallala aquifer is/are true?
   a. It lies beneath the Great Plains and was formed during the last Ice Age
   b. It is the largest aquifer in North America and irrigates about 20% of U.S. croplands
   c. Regular rainfall keeps the aquifer from dropping more than a few feet
   d. A and B above
   e. A and C above

4.   Which body of water was exposed to leaking oil when British Petroleum's Deepwater Horizon rig exploded in the spring of 2010?
   a. Canada's Hudson Bay
   b. Alaska's Prince William Sound
   c. Nunavut's Arctic Circle coastline
   d. The Gulf of Mexico
   e. The Pacific Ocean off the California coast

5.   About 30% of North America's ozone comes from beyond its borders. Which countries are major sources of this and other airborne pollutants?
   a. Canada and Mexico
   b. Germany and Great Britain
   c. Mexico and China
   d. Russia and Japan
   e. Nigeria and Brazil

6.   Where are many producers of acid rain in North America clustered?
       a. New England
       b. The Desert Southwest
       c. The Midwest and Southern Ontario
       d. The Rocky Mountain States and Provinces
       e. The Pacific Northwest

7.   Which of the following cities lies within the original Megalopolis?
       a. Baltimore, Maryland
       b. Kansas City, Missouri
       c. Salt Lake City, Utah
       d. Phoenix, Arizona
       e. Montreal, Quebec

8.   Where were the earliest colonial settlements in North America located?
       a. Along the current U.S. border with Mexico
       b. Around the Great Lakes
       c. In the central states
       d. In the Pacific Northwest region of the U.S. and Canada
       e. Within coastal regions of eastern North America

9.   The most persistent regional migration trend in North America has been movement…
       a. because of pull factors
       b. from the south to the north
       c. from urban areas to rural areas
       d. to Sunbelt states
       e. West

10.   Which of the following statements about African Americans in North America is/are true?
       a. Today, many African Americans are migrating to the South
       b. After emancipation, few former slaves continued to work in the south
       c. At the end of the American colonial period, African Americans made up the majority of the population in southern states of the U.S.
       d. A and B above
       e. A and C above

11.   Today, settlement landscapes of North American cities display the consequences of…
       a. Concentric zones
       b. Counterurbanization
       c. Urban decentralization
       d. A and B above
       e. A, B, and C above

12.   Which of the following is a cultural homeland in Canada?
       a. Acadiana
       b. Black Belt
       c. Hispanic Borderlands
       d. Navajo Reservation
       e. Nunavut

13. Which of the following statements about the relationship between the U.S. and Canada is/are true?
     a. Agricultural competition sometimes creates tension between the two neighbors
     b. Canada and the U.S. joined forces to clean up the Great Lakes
     c. Trade relations between these neighbors is very important
     d. A and B above
     e. A, B, and C above

14. Which part of North America is a leading region of manufacturing exports, specializing in high tech innovations?
     a. Gulf Coast industrial region
     b. Megalopolis
     c. Piedmont manufacturing belt
     d. Silicon Valley
     e. Southern Ontario

15. Which of the following locations in North American is marked by rural poverty?
     a. A ghetto on the west side of Chicago
     b. A retirement community in Florida
     c. A suburb of New York City
     d. Appalachia
     e. British Columbia

Summary of North America

Total Population of North America: About 340 million

Population Indicators for North America

	Highest (country and value)	Lowest (country and value)
Population 2010 (millions)	United States: 309.6	Canada: 34.1
Density per sq km	United States: 32	Canada: 3
RNI	United States: 0.6	Canada: 0.4
TFR	United States: 2.0	Canada: 1.7
Percent Urban	Canada: 80%	United States: 79%
Percent < 15	United States: 20%	Canada: 17%
Percent > 65	Canada: 14%	United States: 13%
Net Migration (per 1000;2000-05)	Canada: 6.3	United States: 3.3

Development Indicators for North America

	Highest (country and value)	Lowest (country and value)
GNI per capita/PPP 2008	United States: $48,430	Canada: $38,710
GDP Avg. Annual Growth (2000-2008)	Canada: 2.5%	United States: 2.4%
Human Development Index (2007)	Canada: 0.966	United States: 0.956
Percent Living below $2/day	----	----
Life Expectancy 2010	Canada: 81	United States: 78
> 5 Mortality 2008	Canada: 6 per 1000	United States: 8 per 1000
Gender Equity 2008	United States: 100	Canada: 99

Chapter 4

Latin America

Learning Objectives

➢ This chapter introduces the concept of Latin America as a region, as well as the countries of Latin America

➢ The chapter emphasizes the various cultural influences of the region, and the impacts of colonialism, immigration, and trade

➢ Emphasis is placed on one of the world's most distinctive physical features, the Amazon rain forest, and discusses the environmental threats associated with this ecosystem

➢ The impacts of globalization are revealed through discussions on the economic disparities exhibited between and within the countries of Latin America

➢ Key concepts introduced in this chapter include altitudinal zonation, grassification, dependency theory, growth poles, and *maquiladora*

Chapter Outline

1. **Introduction**
    A. The modern states of Latin America are multiethnic, with distinct indigenous and immigrant profiles; the states exhibit different rates of social and economic development
        a. There are wide differences in area and population size among the countries in Latin America; significant differences between the wealthy and the impoverished are also well marked
    B. The concept of Latin America as a distinct region has been in existence for approximately one century
        a. Latin America is bound by the Rio Grande (known as the *Rio Bravo* in Mexico) in the north and Tierra del Fuego in the south; the term "Latin America" was coined by French geographers to distinguish the Spanish- and Portuguese-speaking republics of the region from the English-speaking territories
        b. The only connection between "Latin" America and Latin is the presence of Romance languages; the term is widely used in part because it is so vague to be all-inclusive
    C. Through colonialism, immigration, and trade, the forces of globalization are especially pronounced
        a. Economic globalization is currently evidenced by the predominance of neoliberal policies that encourage foreign investment, export production, and privatization; intraregional trade has been encouraged by various trade associations and agreements
        b. Despite growing industrialization and the promotion of export-oriented products, much of the region remains characterized by extractive industries; lumber, natural gas, oil, and copper are significant resources
2. **Environmental Geography: Neotropical Diversity and Urban Degradation**
    A. Most of the region lies within the tropics; however, parts of Argentina and Chile lie south of the Tropic of Capricorn, while much of northern Mexico lies north of the Tropic of Cancer
    B. Because of its vast size and relatively low population density, Latin America has not experienced the level of environmental degradation found in other tropical regions; deforestation, habitat loss, and the loss of biodiversity are significant, and growing, environmental problems
    C. The most significant natural resource challenge is to balance the economic benefits of resource extraction with the ecological soundness of conservation
    D. **Rural Environmental Issues:** Declining Forests and Degraded Farmlands

a. **Forest Destruction and Biodiversity Loss:** deforestation is a significant environmental problem in the Amazon basin and the eastern lowlands of Central America and Mexico; some woodland areas, such as the Atlantic coastal forests of Brazil, the Pacific forests of Central America, the evergreen rain forest of southern Chile, and the coniferous forests of northern Mexico have been especially hard-hit

b. Deforestation has resulted in significant loss of biological diversity; approximately 50 percent of the world's species are found within the rainforests of Latin America, despite this biome accounting for only 6 percent of the world's landmass

   i. Deforestation in Latin America has resulted primarily from cut-and-burn agricultural practices and the cutting of forests to make room for cattle ranges or permanent settlements; with few exceptions, deforestation has not resulted from hardwood extraction for markets

   ii. Brazil has been widely criticized for its Amazon forest policies; during the past four decades, approximately 20 percent of the Brazilian Amazon has been deforested; critics speak of an *arc of deforestation*—a swath of agricultural development and deforestation along the southern edge of the Amazon basin

c. Grassification is the conversion of tropical forest into pasture; this practice has contributed to deforestation and is particularly widespread in southern Mexico, Central America, and the Brazilian Amazon; the pastures are used primarily for cattle ranching; other suitable grasslands exist, such as the Llanos in Colombia and Venezuela, and the Chaco and Pampas in Argentina

d. **Problems on Agricultural Lands:** the modernization of agriculture has produced other environmental problems, including an erosion of genetic diversity as new hybrid varieties of corn, beans, and potatoes have been introduced; the widespread use of chemical fertilizers and pesticides have contaminated local water supplies; farm-workers have been directly exposed through the mishandling of toxic agricultural chemicals; soil erosion and fertility decline has occurred

E. **Urban Environmental Challenges:** Latin American cities exhibit some of the worst environmental problems of the region

a. Urban areas are troubled by air pollution, inadequate supplies of (fresh) water, garbage removal, illegal housing (squatter) settlements

b. The Valley of Mexico epitomizes many of Latin America's urban environmental problems

   i. The Valley lies at the southern end of Mexico's Central Plateau; Mexico City, a metropolitan area of 18 million people, is located in the Valley; the region is beset with problems of air quality, adequate water, and subsidence (soil sinkage) caused by groundwater depletion

F. **Western Mountains and Eastern Shields**

a. Geologically, the Atlantic side of South America is distinguished by large upland plateaus, called *shields*; these shields have historically been the site of the region's most extensive human settlements

b. The Andes mountains are a prominent physical feature of the region; stretching for nearly 5,000 miles from northwestern Venezuela to Tierra del Fuego, the Andes are an active volcanic mountain range; in Colombia the Andes divide into three mountain ranges

   i. The *Altiplano*, a treeless high plain of the Andes, is located where the mountain range traverses Peru and Bolivia; Lake Titicaca and Lake Poopo are located in the Altiplano, as well as numerous mining sites

c. **The Uplands of Mexico and Central America:** this region is the most important in terms of long-term settlement; most major cities of Mexico and Central America are located throughout the Mexican Plateau and the Volcanic Axis of Central America

   i. Vast deposits of silver, copper, and zinc are found throughout the Mexican Plateau

   ii. The Volcanic Axis stretches along the Pacific coast of Central America; many of the 40-plus volcanoes are still active; the rich volcanic soil yields a wide variety of domestic (e.g., corn, beans, squash) and export (e.g., beef, cotton, and coffee) crops; most of Central America's population is located throughout this region

    d. **The Shields**: South America has three major shields: the Brazilian shield, the Guiana shield (discussed in Chapter 5), and the Patagonian shield

        i. The Brazilian shield covers much of Brazil, from the Amazon basin in the north to the Plata basin in the south; it is Latin America's largest and most important region in terms of natural resources and settlement; elevated basins throughout the shield offer mild climates and fertile soils that are conductive to agriculture; the Parana plateau is located on the southern end of the Brazilian shield; it is an exceptionally productive agricultural region

        ii. The Patagonian shield occupies the southern tip of South America; it is sparsely settled and dominated by steppe vegetation; sheep raising and offshore oil production are key economic activities

G. **River Basins and Lowlands**

    a. **Amazon Basin:** world's largest river system by volume and area (drains approximately 2.4 million square miles; second longest river by length; rainfall exceeds 60 inches; exhibits some seasonality in precipitation; watershed shared by eight countries; basin is sparsely settled; development of basin is increasing rapidly

    b. **Plata Basin:** region's second largest basin; includes the Parana, the Paraguay, and the Uruguay rivers; much of the Plata is economically productive through large-scale mechanized agriculture

    c. **Orinoco Basin:** third largest river basin in region; dominated by tropical grassland called the *Llanos*; sparsely settled; limited development aside from large cattle ranches; petroleum production is significant

H. **Climate Patterns**

    a. Tropical climate with little variation in temperatures; seasonality in precipitation is apparent

        i. **Tropical humid:** lowlands, especially east of the Andes, dominated by savannas or forests

        ii. **Desert:** found along Pacific coasts of Peru and Chile, in Patagonia, northern Mexico, and Bahia of Brazil

        iii. **Mid-latitude:** found in Argentina and Uruguay, as well as parts of Paraguay and Chile

    b. **Altitudinal Zonation:** declining temperatures with increased elevation contributes to distinct ecosystems and agricultural potential; zones include *tierra caliente* (hot land, ranges from sea level to 3,000 feet; supports sugar cane, tropical fruits, livestock); *tierra templada* (temperate land, ranges from 3,000 to 6,000 feet; supports coffee, maize, warm-weather vegetables, cut flowers, short-horn cattle); *tierra fria* (cold land, ranges from 6,000 to 12,000 feet; supports wheat, barley, maize, tubers, sheep, llama, alpaca, and guinea pigs); and *tierra helada* (frozen land; ranges from 12,000 feet and higher; supports potato and quinoa)

    c. **El Niño:** refers to an oscillation of the ocean-atmosphere system in the tropical Pacific; normally, Pacific coastal waters of Ecuador and Peru are dominated by cold ocean currents; during El Niño year, warm current arrives; significantly impacts and alters normal weather patterns, causing floods in some places and droughts in others

    d. **Latin America and Global Warming:** global warming has immediate and long-term implications for the region; in short-term, climate change will influence agricultural productivity, water availability, changes in ecosystems, and incidence of diseases; in the long-term, impacts are less clear, but climates will be altered; sea-level rise less of concern in this region compared to other, more vulnerable regions such as the Caribbean or Oceania; highland areas perhaps most vulnerable

3. **Population and Settlement**

A. **Settlement Patterns:** Latin America does not have tradition of great river basin civilizations; major population clusters located in interior plateaus and valleys of Central America and Mexico, while interior lowlands of South America sparsely settled; settlement in South America historically in highland regions and coastal regions

B. Rapid population growth in 1960s and 1970s, with slowing growth during 1980s and 1990s; declining fertility rates; increasing life expectancies

C. **The Latin American City:** region is highly urbanized; approximately 75 percent of population living in cities; Argentina, Chile, Uruguay, and Venezuela are more than 90 percent urbanized

    a. Region exhibits high levels of urban primacy, a condition in which a country has a primate city three to four times larger than any other city; examples include Lima, Caracas, Guatemala City, Santiago, Buenos Aires, and Mexico City; primacy usually considered a liability because of over-concentration of services; some governments have built new cities to offset primacy, examples include Ciudad Guayana and Brasilia

    b. Three emerging *megalopolises* include Mexico City-Puebla-Toluca-Cuernavaca on Mesa Central; Niteroi-Rio de Janeiro-Santos-Sao Paulo-Campinas in southern Brazil; and Rosario-Buenos Aires-Montevideo-San Nicolas in Argentina and Uruguay's Rio Plata basin

    c. **Urban Form:** distinct urban morphology based on colonial origins and contemporary growth (Figure 4.17); composed of central business district, elite spine, concentric residential sectors, peripheral squatter settlements—impoverished areas with limited services and basic infrastructures

        i. Informal sector: economic sector that relies on self-employed, low-wage jobs that are usually unregulated and untaxed; predominate within squatter settlements

    d. **Rural-to-Urban Migration**: many rural residents, having become displaced, are moving to potential opportunities of cities; source of much urban growth

D. **Patterns of Rural Settlement:** continued importance of peasant subsistence farming, but recent increase in mechanized, capital-intensive farming; rural areas becoming less isolated to urban locations; international migration contributing to globalization of rural villages

    a. **Rural landholdings:** a system of large estates (*encomienda*), owned by absentee landlords and use of hired and slave labor; *latifundia*, an entrenched pattern of large estates associated with encomienda; *minifundia*, pattern associated with peasants farming small plots for subsistence; agrarian reform: policies of redistribution of land to peasants—often associated with conflict and violence

    b. **Agricultural Frontiers:** opening of new land to peasants; serves also to exploit natural resources and to strengthen political boundaries; examples include Brazil's Trans-Amazon Highway, Peru's Carretera Marginal (Marginal Highway), Mexico's Tehuantepec, Guatemala's northern Peten

E. **Population Growth and Movement:** high rates of growth during 1960s and 1970s, resultant from natural increase and immigration; reduced growth since 1980s because of declining fertility

    a. **European migration:** many Latin American governments encouraged immigrants from Europe throughout 19th century; southern cone of Latin America (Argentina, Chile, Uruguay, and southern Brazil) received many European migrants, mostly from Italy, Portugal, Spain, and Germany

    b. **Asian migration:** large numbers of migrants from China and Japan arrived in 19th century as laborers; currently many migrants from South Korea, engaged in small businesses

    c. **Latino Migration and Hemisphere Change:** considerable interregional migration resultant from changes in economic conditions of countries; political turmoil also influences both interregional and international migration patterns; Mexico is largest source country of Latino immigrants into the United States; overall Latin America is region of emigration, with many families reliant upon *remittances* (monies sent back home); immigrants continue to maintain close contact with origins, a phenomenon termed *transnationalism*

4. **Patterns of Cultural Coherence and Diversity**

A. Many complex civilizations existed in region prior to European arrival (1492); examples include the Mayan, Aztec, and Incan civilizations

B. European colonization was syncretic process as indigenous groups often incorporated into Spanish or Portuguese empires; some Amerindian cultures were resilient; others largely destroyed through forced assimilation

C. **Demographic Collapse:** term used in reference to dramatic decline in indigenous population as a result of contact with Europeans; approximately 90 percent of region's pre-1500 population of 54 million inhabitants died because of introduced diseases, war, forced labor, and starvation

D. **The Columbian Exchange:** an immense biological "swap" that occurred from the 16th century onwards; European agricultural products such as wheat, olives, grapes, pigs, cows, horses, sugar cane, and coffee diffused into the region; Amerindian crops including maize/corn, potatoes, hot peppers, tomatoes, pineapples, cacao, and avocados were diffused to Europe

   a. The movement of Old World animals dramatically impacted the Americas; swine, sheep, cows, and horses became widely used for staple products and as draft animals

E. **Indian Survival:** largest indigenous populations found in Mexico, Guatemala, Ecuador, Peru, and Bolivia; land remains key factor for overall cultural survival; many indigenous groups are demanding formal political and territorial recognition

F. **Patterns of Ethnicity and Culture:** considerable European influences, but mostly a complex ethnic mix evolved; racial mining was prevalent, leading to system of racial castes, e.g., *mestizo* (resulting from union of Spanish and Indian persons), *castizo* (resulting from union of mestizo and Spanish women), *mulattoes* (progeny of European and African unions), and *zambos* (offspring of Africans and Indians); these divisions later replaced by four-fold racial classification of *blanco* (European ancestry), *mestizo* (mixed ancestry); *indio* (Indian ancestry); and *negro* (African ancestry)

G. **Enduring Amerindian Languages:** About two-thirds of Latin Americans speak Spanish; one-third speak Portuguese; important pockets of indigenous languages remain, especially in Central Andes, southern Mexico, and Guatemala

H. **Blended Religions:** approximately 90 percent of the region's population is Catholic; others practice *syncretic religions*, which are blends of different belief systems

I. **Global Reach of Latino Culture:** Many aspects of Latino culture have been absorbed into globalizing world culture, e.g., tango, soccer; many Latino singers have attained global popularity, including Shakira and Ricky Martin; *telenovelas*, popular nightly soap operas, have been exported globally

   a. **National Identities:** country's within the region exhibit strong national identities; identities, however, are fluid, context-dependent, and evolving

5. **Geopolitical Framework**

A. **Post-European contact:** territories of Latin America were divided by European colonial powers; by 19th century, many sovereign states emerged, but territorial boundaries reflected colonial legacy

B. **Organization of American States:** regional organization chartered in 1948; promoted a neutral Pan-American vision of hemispheric cooperation; reaction against political and economic dominance exercised by the United States

C. **Iberian Conquest and Territorial Division**

   a. *Treaty of Tordesillas:* Papal charter in 1493–1494 that led to division of Latin American into Spanish (western) and Portuguese (eastern) spheres

      i. **Portuguese:** initially little interest in Brazil; development of extensive sugar estates and slave trade after 16th century

      ii. **Spanish:** aggressive conquest and settlement of its territories; development of silver mines; promotion of agriculture; manufacturing was largely forbidden

D. **Revolution and Independence:** Spanish authority challenged in 19th century; many revolutionary movements between 1810 and 1826; most modern states of Latin America formed during this period

E. **Persistent Border Conflicts:** territories were not clearly demarcated, leading to continued conflicts throughout 19th and 20th centuries

F. **Trend Toward Democracy:** Although most countries have been independent for nearly 200 years, political instability has been a defining feature; since 1980s, however, trend has been toward democratically elected governments and promotion of free markets; many countries exhibit ongoing popular frustration with falling incomes, rising violence, corruption, and chronic underemployment

G. **Regional Organizations:** *supranational organizations* (governing bodies that include several states) and *subnational organizations* (groups that represent areas or peoples within a state) are affected political developments in Latin America

    a. **Trade Blocs:** regional alliances that foster internal markets and reduce trade barriers; examples include the Latin American Free Trade Association (LAFTA), the Central American Common Market, and the Andean Community; in the 1990s Mercosur (formed originally with Brazil, Argentina, Uruguay, and Paraguay) and the North American Free Trade Agreement (composed of Mexico, the United States, and Canada) were established; in 2004 the Central American Free Trade Agreement was signed by the Dominican Republic, Nicaragua, and Costa Rica; combined, these trade blocs and agreement significantly impact the political economy of the region

H. **Insurgencies and Drug Trafficking:** political insurgencies and the violence associated with drug trafficking characterize many countries in the region; the United States has been actively involved in counternarcotics operations in Central and South America

    a. **Indigenous Groups:** many indigenous organizations have been formed to protest neoliberal policies and globalization; the involvement of these groups reflects both a deepening of democracy and the emergence of ethnically driven politics

6. **Economic and Social Development**: most Latin American countries are classified as middle-income; economic disparities between and within states remain; Brazil and Mexico are the economic engines of the region; Bolivia is among the region's most impoverished states

A. **Development Strategies:** beginning in the 1960s, multilateral agencies such as the Intra-American Development Bank and the World Bank promoted large-scale development projects, green revolution technologies, and other modernization programs; many countries in the region exhibited more economic diversification; in the 1990s, most Latin American governments adopted free-market policy reforms, termed the *Washington Consensus*; many of these neoliberal reforms were highly unpopular, leading to political and economic turmoil

B. **Industrialization:** since the 1960s, most governments have emphasized manufacturing; considerable growth has been fueled by investment from foreign companies

C. **Maquiladoras and Foreign Investment:** maquiladoras are assembly plants located in Mexico; these factories manufacture an assortment of products, including automobiles, consumer electronics, and apparel; while providing employment, these have been critiqued because of low wages, high labor turnover, lax health and safety regulations, and lax environmental regulations

    a. **Border Industrialization:** the emergence of maquiladoras associated with border industrialization program initiated in the 1960s; promotion of export-oriented products, heavily dependent upon foreign investment

D. **Entrenched Informal Sector:** informal activities include both legal (e.g., street vending, transportation services, garbage picking/scavenging, and street-performing), and illegal (e.g., drug trafficking, prostitution, and money laundering) activities; many economies of the region are heavily dependent on informal activities; the widespread existence of the informal economy is an indicator of Latin America's poverty

E. **Primary Export Dependency:** dependence on the export of raw materials, including minerals and agricultural products; many countries are heavily dependent on one or two export products; between the 1980s and 1990s, many countries diversified their economies; since the 1990s, high demand for resources in Asia has led to a boom in the primary export of commodities

    a. **Agricultural Production:** overall trend has been to diversify and mechanize agriculture; prime example is the Plata Basin and the production of soybeans; mechanized agriculture is contributing to the elimination of forest and savanna and is pushing out subsistence-based peasant farmers

    b. **Mining and Forestry:** extraction of silver, zinc, copper, iron ore, bauxite, and gold is economic mainstay of region; mining is mostly becoming more capital-intensive and less labor-intensive (gold mining is exception); logging has led to many environmental problems; plantation forests are increasing, contributing to process of deforestation of native forests

    c. **Energy Sector:** Mexico, Venezuela, and Colombia are important oil producers; Venezuela and Bolivia have significant reserves of natural gas; Brazil has invested in ethanol production

F. **Latin America in the Global Economy**

    a. ***Dependency Theory:*** theory developed in 1960s; premise is that expansion of European capitalism created underdevelopment of region; "peripheral" countries became impoverished through their dependence on the prospering "core" countries; through unequal trade relations and patterns, dependent countries were highly vulnerable to fluctuations of global economy

        i. Adherents to dependency theory promoted self-sufficiency, growth of internal markets, agrarian reform, and vigorous state intervention

        ii. Many Latin American states are still dependent upon foreign countries, especially the United States; both trade and foreign investment are important connections

    b. Latin America linked to global economy through immigration and remittances

    c. **Neoliberalism as Globalization:** *neoliberal policies* stress privatization, export production, direct foreign investment, and minimal restrictions on imports; neoliberalism epitomizes globalization by turning away from self-sufficiency and state intervention; many Latin American governments have embraced neoliberal policies; Chile is especially notable

    d. ***Dollarization:*** a process by which a country adopts, in whole or in part, the U.S. dollar as its official currency; in 1904 Panama dollarized its economy; Ecuador dollarized its economy in 2000, and El Salvador followed in 2001; this process is supposed to address currency devaluation and hyperinflation; a country that adopts dollarization no longer controls its monetary policy and becomes dependent on the U.S. Federal Reserve; may help in short-term, but is not a popular policy

G. **Social Development:** marked improvements in life expectancy, child survival, and educational quality; both government policies and grassroots and nongovernmental organizations have contributed to improvements in social well-being; examples include Brazil's poverty reduction program

    a. Significant spatial variation in well-being are evident; all countries have spatial inequities regarding income and availability of services

    b. **Race and Inequality:** both ethnicity and race, particularly in Mexico and Brazil, augment economic and social inequalities; throughout Latin America, Amerindians and persons of African ancestry are among the most economically and socially marginalized

    c. An emerging middle class, with its own aspirations, is evident

    d. **The status of women:** contradictions exist; many Latinas find employment in the labor market, participate in politics, and have access to education and family planning; however, patriarchal tendencies remain, and women's participation in economic, social, and political activities remain significantly lower than men's participation

Summary

- Latin America (and the Caribbean) was the first world region to be colonized by Europe; in the process, approximately 90 percent of the indigenous population died; the slow demographic recovery, marked by considerable immigration, has resulted in a complex society of ethnic and cultural hybridity
- Unlike most other developing regions, Latin America is highly urbanized; cities combine elements of both formal and informal economies
- Compared to Europe and Asia, this region remains well-endowed with natural resources; environmental problems, such as deforestation, are substantial
- Uneven development and economic frustration have contributed to emigration
- Latin American governments have widely adopted neoliberal policies; some governments have faltered in their attempts and popular protests have resulted; new political actors, including indigenous groups and women, have emerged

Research or Term Paper Ideas

- What were the initial goals set forth when the North American Free Trade Agreement (NAFTA) was signed? Were these goals met? How have the participating countries progressed since the establishment of NAFTA?

- Research the physical and political geography of the Amazon rainforest. Where is it located? How would you describe its ecology? What specific factors have led to the deforestation of the Amazon? How many species have been lost? What policies have been introduced to minimize the environmental degradation associated with the deforestation of the Amazon? Have these been successful?

- In recent years the on-going 'war on drugs' has overlapped with the 'war on terror'. How have foreign governments, such as the United States, changed policies in response to the wars on drugs and terror? How have these policies impacted the relevant countries of Latin America?

- Latin American culture is becoming more global. This is seen in the diffusion of music, dance, film, and literature, to name but a few. Select one topic, such as music, and discuss specific ways in which this cultural feature has significantly impacted other regions.

Practice Quizzes

Answers appear at the end of this book

Vocabulary Matching: *Match the term to its definition.*

A. Agrarian reform
B. Altiplano
C. Altitudinal zonation
D. Columbian exchange
E. Dependency theory
F. El Niño
G. Environmental lapse rate
H. Fair trade

I. Grassification
J. Informal sector
K. Maquiladora
L. Megacity
M. Mercosur
N. Neoliberal policies
O. Remittances
P. Urban primacy

1. _____ Urban settlement (city) of more than 10 million people.

2. _____ An exchange of people, diseases, plants, and animals between the Americas (New World) and Europe/Africa (Old World) initiated by the arrival of European colonizers in 1492.

3. _____ The conversion of tropical forest into pasture.

4. _____ The largest intermontaine plateau in the Andes; it straddles Peru and Bolivia and ranges in elevation from 10,000 to 13,000 feet.

5. _____ The relationship between higher elevations, cooler temperatures, and changes in vegetation that result from the environmental lapse rate, which in turn results in different bioregions depending on elevation.

6. _____ An abnormally warm Pacific current that usually arrives along coastal Ecuador and Peru in December; it may bring torrential rains and floods along the Pacific coast, and drought conditions to interior regions of the Americas.

7. _____ A condition in which a country has a single large city three to four times larger than any other city in the country; in Latin America, Lima, Caracas, and Mexico City are examples.

8. _____ The drop in temperature as one ascends higher in the atmosphere; on average, the temperature declines 3.5 degrees Fahrenheit for every 1,000 feet.

9. _____ Money sent by immigrants to their country of origin to support family members left behind.

10. _____ Assembly plants on the Mexican border built by foreign capital; most of their products are exported to the U.S.

11. _____ Economic sector that relies on self-employed, low-wage jobs (such as street-vending, shoe-shining, artisan crafts) that are virtually unregulated and untaxed.

12. _____   The Southern Common Market established in 1991 that calls for free trade among member states and common external tariffs for nonmember states; Argentina, Paraguay, Brazil, and Uruguay are members; Chile is an associate member.

13. _____   A popular theory to explain patterns of economic development in Latin America; its central premise is that underdevelopment was created by the expansion of European capitalism into the region that served to develop core countries in Europe and impoverish and make dependent peripheral areas such as Latin America.

14. _____   Redistribution of land, usually to provide farming opportunities to peasants.

15. _____   Economic policies that emphasize export production, privatization, oppose restrictions on imports, and encourage direct foreign investment.

Multiple Choice: *Choose the word or phrase that best answers the question*

1.   What are the boundaries of Latin America?
    a. Amazon rain forest in the north to Montevideo in the south
    b. Llanos in the north to Patagonia in the south
    c. Nicaragua in the north to Bahia Blanca in the south
    d. Rio Grande in the north to Tierra del Fuego in the south
    e. Yucatan Peninsula in the north to La Plata in the south

2.   In several countries of Latin America, there is still a strong Indian presence. Which of the following countries has the weakest Amerindian presence?
    a. Bolivia
    b. Ecuador
    c. Guatemala
    d. Peru
    e. Uruguay

3.   Latin America has not experienced the same level of environmental degradation as North America, East Asia, and Europe. What is the reason?
    a. Relatively low population density
    b. Large size
    c. A long-enduring respect for the environment
    d. A and B above
    e. B and C above

4.   Of all the environmental problems in Latin America, which one is the most critical for biological diversity?
    a. Air pollution
    b. Loss of tropical rain forest
    c. Smog
    d. Subsidence
    e. Water quality

5.   What have been the results of the use of farm chemicals (pesticides, herbicides, fertilizers) in Latin America?
      a. Contamination of groundwater supply
      b. Rashes and burns on the skin of farm workers
      c. Rise in serious birth defects
      d. A and B above
      e. A, B, and C above

6.   Which of the following statements about the Andes Mountains are true?
      a. They are relatively young and volcanoes and earthquakes are common there
      b. They are nearly 5000 miles long and include rich veins of precious metals and minerals
      c. The Andes are the world's tallest mountains
      d. A and B
      e. A, B, and C

7.   What is the key element that causes altitudinal zonation?
      a. Wind caused by monsoons
      b. Variations in precipitation caused by orographic precipitation
      c. Temperature drop known as environmental lapse rate
      d. Storms caused by El Niño
      e. All of the above

8.   What weather pattern signals the arrival of El Niño?
      a. Unseasonably cool temperatures
      b. Torrential rains
      c. Tornadoes
      d. Monsoonal winds
      e. Drought

9.   What is the country of origin of the largest number of legal immigrants to the U.S.?
      a. Venezuela
      b. Argentina
      c. Nicaragua
      d. Mexico
      e. Panama

10.   In which of the following countries of Latin America is there no urban primacy?
      a. Brazil
      b. Guatemala
      c. Mexico
      d. Peru
      e. Venezuela

11.   What was the basis for political and economic power in Latin America since the colonial era?
      a. Military weapons
      b. Grass-roots organization
      c. Education
      d. Control of land
      e. All of the above

12.   From which countries did migrants come to Latin America to work in Brazil and Peru?
      a.   North and South Korea
      b.   Libya and Egypt
      c.   Indonesia and Malaysia
      d.   India and Pakistan
      e.   China and Japan

13.   What is the key factor that unifies Latin America as a region?
      a.   Climate
      b.   Cultural unity
      c.   History of colonization
      d.   The Amazon Basin
      e.   The Andes Mountains

14.   Which country of Latin America was at the center of the drug trade when it began in the 1970s?
      a.   Argentina
      b.   Brazil
      c.   Chile
      d.   Colombia
      e.   Mexico

15.   What is a maquiladora?
      a.   A squatter settlement in Latin America
      b.   Mexican assembly plants that line the border with the U.S.
      c.   A small plot of land provided to peasants in Latin America for their subsistence
      d.   A person of European and Indian ancestry
      e.   A cultural trait ascribed to women in Latin America

Summary of Latin America

Total Population of Latin America: About 542 million

Population Indicators for Latin America

	Highest (country and value)	Region Average	Lowest (country and value)
Population 2010 (millions)	Brazil: 193.3	31.89	Uruguay: 3.4
Density per sq km	El Salvador: 294	57.82	Bolivia: 9
RNI	Guatemala: 2.8	1.54	Uruguay: 0.5
TFR	Guatemala: 4.4	2.54	Chile; Costa Rica: 1.9
Percent Urban	Uruguay: 94%	70.53%	Guatemala: 47%
Percent < 15	Guatemala: 42%	30.76%	Costa Rica; Uruguay: 23%
Percent > 65	Uruguay: 14%	6.47%	Nicaragua: 3%
Net Migration (per 1000;2000-05)	Costa Rica: 1.3	-2.37	El Salvador: -9.1

Development Indicators for Latin America

	Highest (country and value)	Region Average	Lowest (country and value)
GNI per capita/PPP 2008	Mexico: $14,340	$8,900	Nicaragua: $2,620
GDP Avg. Annual Growth (2000-2008)	Panama: 6.6%	4.5%	Mexico: 2.7%
Human Development Index (2007)	Chile: 0.878	0.800	Nicaragua: 0.699
Percent Living below $2/day	Chile: 2.4%	15%	Nicaragua: 31.8%
Life Expectancy 2010	Chile; Costa Rica: 79	74	Bolivia: 66
< 5 Mortality 2008	Chile: 9 per 1000	23 per 1000	Bolivia: 54 per 1000
Gender Equity 2008	Argentina; Columbia: 104	101	Guatemala: 94

Chapter 5

The Caribbean

Learning Objectives

➤ This chapter introduces the Caribbean as a unique cultural and physical region
➤ The chapter develops the concept of plantation agriculture; understanding this form of agriculture will help you to understand agricultural transformations in other regions
➤ The impact of colonialism on contemporary geopolitical events is highlighted
➤ Tourism as an economic development activity is emphasized
➤ The understanding of globalization is strengthened through introduction of multilateral practices such as off-shore assembly plants and banking

Chapter Outline

1. **Introduction**
   A. The Caribbean was the first region of the Americas to be colonized by the Europeans
   B. The region's identity is unclear; it is often considered both part of and separate from Latin America
      a. The region is home to 43 million inhabitants in 26 countries and territories
      b. Countries and territories in the region exhibit a broad range of sizes; the British dependency of the Turks and Caicos has 12,000 people while the island of Hispaniola is home to 20 million
      c. Three states on the mainland of South America (Guyana, Suriname, and French Guiana) and one state in Central America (Belize) are considered part of the Caribbean
   C. From the 16th century through the 19th century European colonial powers rivaled for control of the region; since the 19th century, the United States has been the major geopolitical force in the region
   D. Agriculture (especially plantations) remains a dominant economic activity; tourism, offshore banking, and manufacturing are also important
   E. The region is characterized by cultural diversity, high population densities, and vast disparities in wealth
   F. The concept of *isolated proximity* has been proposed to explain the region's unusual and contradictory position in the world; the region's isolation accounts for its cultural diversity and limited economic opportunities, while its proximity to North America ensures its transnational connections and economic dependency
2. **Environmental Geography**
   A. **Environmental Issues:** the region's ecology has been significantly transformed as a result of colonization and global trade; the depletion of biological resources has been especially severe
   B. **Deforestation on the Islands:** Prior to the arrivals of the Europeans, much of the region was covered in tropical rain forests and deciduous forests; most were cleared for plantations and the need for fuel and lumber
   C. **Managing Rimland Forests:** The Caribbean *rimland* includes the coastal zone of the mainland, stretching from Belize through Central America and the northern coast of South America
      a. The biological diversity and stability of the rimland states are less threatened than the islands; public awareness of conservation efforts is greater; many protected areas have been established
      b. Some areas, including Guyana and Suriname, remain battlegrounds between conservationists, indigenous peoples, and developers

D. **Urban Environmental Problems:** cities in the region confront problems stemming from water quality and waste disposal; for many of the urban poor, water and sewage services are overtaxed or nonexistent
   a. Most island economies cannot afford to improve basic urban infrastructure
   b. Freshwater supplies fall short of domestic needs
E. **Islands and Rimland Landscapes:** The Caribbean Sea lies between Antillean islands and mainland Central and South America; the region is interconnected by trade; abundant in marine diversity, but not in quantities of species, thus not supportive of large-scale commercial fishing; the Antillean islands are divided into two groups
   a. **Greater Antilles:** composed of the islands of Cuba, Jamaica, Hispaniola (shared by Haiti and the Dominican Republic), and Puerto Rico; home to most of the Caribbean's population; contains significant arable lands; farming is important but soils are nutrient poor, heavily leached, and acidic; high mountains characterize physical geography
   b. **Lesser Antilles:** double arc of small islands, stretch from Virgin Islands to Trinidad and from St. Kitts to Grenada; smaller in size and population than Greater Antilles; historically were important footholds for European colonial powers; volcanic in origin
   c. **The Rimland:** composed of Belize and the Guianas; retain significant amounts of forest cover; agricultural patterns heavily influenced by geology and soils
F. **Climate and Vegetation:** tropical climate, with warm temperatures and abundant rainfall year round; seasonality marked by changes in precipitation amounts rather than temperature; dominance of tropical forest vegetation; some grasslands
   a. **Hurricanes:** hurricane zone encompasses the Caribbean; approximately six hurricanes impact the region each year
   b. **Forests, Savannas, and Mangroves:** tropical forests, once widespread, now largely confined to rimland; small pockets remain elsewhere, usually surrounded by fields and savannas; coastal mangroves are poorly suited for human settlement but key ecosystem for marine life; removal of mangroves has resulted in increased erosion and reduction in marine habitat
   c. **Caribbean and Global Warming:** effects of global warming on region have been sea-level rise, increased intensity of storms, variable rainfall leading to floods and droughts, loss of biodiversity; region is not a major contributor of greenhouse gases

3. **Population and Settlement**
   A. **Generally high population densities;** increasingly urbanized; but region exhibits significant variation—islands of Greater Antilles more populated, while Lesser Antilles less populated, but still can have high population densities because of smaller territories; access to arable land a major problem for many islands; mainland states are sparsely populated
   B. **Demographic Trends:** throughout 17th through 19th century mortality rates of indigenous peoples were high because of disease, maltreatment, and malnutrition; large-scale importation of slaves from Africa—who also suffered from high mortality; from mid-19th century, gradual improvement in sanitary conditions; natural population increase, with peak periods of growth from 1950s and 1960s; growth rates have declined somewhat and stabilized
      a. **Fertility Decline:** most significant demographic trend; educational improvements, urbanization, and preference for small family size have contributed to fertility declines
      b. **Rise of HIV/AIDs:** Although in decline, the rate of infection remains twice that of North America (but considerably less than that of sub-Saharan Africa)
         a. AIDs is largest cause of death among young men in English-speaking Caribbean
         b. Haiti has been highly impacted
         c. Many countries have increased educational awareness to prevent spread of disease
   C. **Emigration:** emigration to other parts of the Caribbean, North America, and Europe began in 1950s and has continued; largely economic driven; often referred to as *Caribbean diaspora*
      a. Intraregional movements also significant
      b. Circular migration—individual seeks employment in foreign country, returns home, perhaps multiple times; very common pattern

    c.   Chain migration—one family member moves to new country and over time brings relatives and friends; can account for formation of immigrant enclaves in destination countries

    d.   Transnational migration—the straddling of livelihoods and households between two countries; this is an increasingly common practice; those from Dominican Republic are considered the most transnational

  D. **Rural–Urban Continuum:** structure of Caribbean communities reflects plantation and slave legacy; rural communities tend to be loosely organized; labor is transient; small farms scattered on available land

  E. **Caribbean Cities:** rural–urban migration has increased since 1960s because of mechanization of agriculture, offshore industrialization, and rapid population growth; cities have grown in size

    a.   Spatial variation in levels of urbanization; Cuba is most urbanized; Haiti is the least

    b.   Layout of cities often reflects colonial legacy; cultural diversity reflected

    c.   **Housing:** shortage of urban employment opportunities and housing resulted in rise in squatter settlements in most cities; cities in Cuba are exception, because of housing developments provided by socialist government

4. **Cultural Coherence and Diversity**

  A. Common historical and social processes unite the region, whose people have many linguistic, religious, and ethnic differences; process of *creolization*—the blending of different cultures—is very apparent throughout region

  B. **Cultural Imprint of Colonialism:** European colonialism imposed different social systems and cultures; colonial brutality, enslavement, warfare, and introduction of new diseases led to massive depopulation of indigenous Amerindian peoples; slave labor from Africa was used

  C. **Plantation America:** term used to designate cultural region extending from midway up coast of Brazil through the Guianas and Caribbean into the southeastern United States; plantation agriculture is production system based on monocrop production (a single commodity, such as sugar); plantation societies were created along class lines and racial divisions

  D. **Asian Immigration:** from mid-19th century, many Asians (mostly from South and Southeast Asia) recruited to work in region as *indentured labor* (workers contracted to labor on estates for a set period of time); many remained and intermarried; especially important migration system to Suriname, Guyana, and Trinidad and Tobago

  E. **Creating a Neo-Africa:** African slaves first brought to region in 16th century; continued through 19th century; forced migration was part of broader African diaspora associated with the trans-Atlantic slave trade

    a.   **Maroon Societies:** communities of runaway (escaped) slaves, called *maroons*; surviving Maroon settlements protected African traditions, including farming practices, house designs, community organization, and language; many of these communities remain, but are threatened by globalization, especially the construction of dams, mining operations, and logging

    b.   **African Religions:** linked to Maroon societies, but more widely diffused; evolved into new unique patterns of worship; most widely practiced are Voodoo in Haiti, Santeria in Cuba, and Obeah in Jamaica

  F. **Creolization and Caribbean Identity:** *Creolization*—the blending of African, European, and some Amerindian cultural elements into unique sociocultural systems found in Caribbean; Creole identities are complex; illustrate cultural and natural identities; expressed in language, religion, arts, literature, and music

    a.   **Language:** Dominant languages in region are European (Spanish, French, English, and Dutch); new "Creole" languages have emerged to become the *lingua franca*; in 1960s Creole languages became politically and culturally charged with national meaning

    b.   **Music:** roots of modern Caribbean music reflect a combination of African rhythms with European forms of melody and verse; region reveals many distinct local musical forms

    c.   **Baseball to Beisbol:** United States' influence has led to popularity of baseball; many players in leagues around the world, but especially in North America, originate from the Caribbean

5. **Geopolitical Framework**: Caribbean colonial history is patchwork of rival powers fighting over profitable tropical territories; many territories have changed colonial possession repeatedly; by 19th century, influence of United States became more apparent; the U.S. promoted the *Monroe Doctrine*, which claimed that the U.S. would not tolerate European military involvement in the Western Hemisphere; today, *neocolonialism* is prevalent throughout region

   A. **Life in the "American Backyard"**: The Caribbean was commonly referred to as the "American Backyard" in early 20th century; U.S. influence on region remains considerable; foreign policy objects of the U.S. were to remove European authority, foster democratic governance, and gain access to resources and markets; U.S. military force was often used

   B. **Commonwealth of Puerto Rico**: Puerto Rico is a commonwealth of the United States; various independence movements seek to secede from the U.S.; Puerto Rico depends on U.S. investment and welfare programs; no legal restrictions regarding migration between Puerto Rico and the U.S.; in 1950s "Operation Bootstrap" was launched—a major industrialization program that transformed Puerto Rico from agrarian economy to an industrial economy

   C. **Cuba and Regional Politics**: in 1950s Fidel Castro led successful socialist revolution in Cuba; country remains the main challenge to U.S. authority in the region; in 1950s and 1960s Castro nationalized all industries and state took possession of foreign-owned properties; the U.S. initiated trade embargoes which remain; under Castro, literacy and public health have improved

   D. **Independence and Integration**:
      a. **Independence Movements**: Beginning in 19th century, colonies began to seek (and achieve) independence; by 1960s and 1970s, many had become independent; many sovereign states unable to meet basic needs of citizens
      b. **Present-day Colonies**: Britain still maintains several crown colonies; France retains "departments" in region; Dutch oversees "autonomous countries" that are part of the Kingdom of the Netherlands; some Caribbean territories maintain their colonial status as an economic asset
      c. **Regional Integration**: most difficult task is to promote economic integration; scattered islands, divided rimland, different languages, and limited economic resources hinder formation of regional trade bloc; the Caribbean Community and Common Market (CARICOM) was formed in 1972—presently has 15 full-member states; CARICOM has produced limited improvements in intraregional trade

6. **Economic and Social Development**: collectively, region is relatively impoverished, though economically better off than much of sub-Saharan Africa, South Asia, and China; most countries are lower-middle income; social gains in education, health, and life expectancy have been significant

   A. **From Fields to Factories and Resorts**: agriculture is declining as source of economic development; reflect shift from monocrop dependence

   B. **Sugar and Coffee**: economic history of region tied to sugar; Caribbean still major exporter of sugar; economic importance in decline because of competition from corn and sugar beets grown in mid-latitudes; coffee is significant export crop, though hindered by instability of global coffee prices

   C. **The Banana Wars**: Latin America is major producer and exporter of bananas; some Caribbean states dependent on export of bananas; production for export has fostered greater economic and social development

   D. **Assembly-plant Industrialization**: since 1950s, foreign companies invited to build factories in region; creation of *free trade zones* (FTZs)—these are duty-free and tax-exempt industrial parks for foreign corporations; controversial—new jobs are created, national economies have diversified, but overall foreign investors gain more than host countries

   E. **Offshore Banking and Online Gambling**: beginning in 1920s, some Caribbean islands have established *offshore banking* centers, which appeal to foreign banks and corporations by offering specialized services that are confidential and tax-exempt; the Bahamas has been the forerunner of this activity; different islands offer specialized and unique services; centers also attract money tied to drug trade and terrorist groups (through money laundering)

a. **Online Gambling:** newest industry in the region—Antigua and St. Kitts are leaders; established in 1999; able to take advantage of cyberspace—online gambling illegal in the United States, but U.S. residents can gamble online in cyberspace

F. **Tourism:** major economic activity, facilitated by environmental, locational, and economic factors

   a. In 1950s Cuba was major tourist destination; more recently, Puerto Rico, the Dominican Republic, Cuba, the Bahamas, Jamaica, and the British Virgin Islands are most visited

   b. Tourism is important source of income, but controversial; not consistent, because subject to overall health of global economy and current political affairs; problem of *capital leakage*— the huge gap between gross receipts and total tourist dollars that remain in Caribbean; most residents of Caribbean do not benefit; environmental degradation is significant but somewhat less destructive problem

G. **Social Development:** measures of social development are generally strong; literacy levels and life expectancy have increased; remittances from migrants thought to help overall level of social and economic development; however, many inhabitants chronically underemployed, poorly housed, and perhaps overly dependent on foreign remittances

   a. **Status of Women:** region is generally matriarchal; existence of strong and self-sufficient female networks; women often have local power, but lag behind in overall status and position; in rural areas, women often excluded from cash economy; female employment in assembly plants is significant

   b. **Education:** many Caribbean states have excelled in promoting education; education is expensive but considered essential for development; many states suffer from *brain drain*— the emigration of trained professionals

H. **Labor-Related Migration:** since mid-20th century, considerable emigration; has impacted community and household structures, regional economies, and overall state economic growth and development; *remittances* (monies sent back home) have been crucial

Summary

- Caribbean is more integrated into global economy than other areas of the developing world; but still remains on economic periphery
- Tropical region has been exploited to produce export commodities, including sugar, coffee, and bananas; serious problems with deforestation, soil erosion, and water contamination
- Population growth has slowed in recent decades; life expectancy and literary rates are quite high
- Caribbean was forged through European colonialism and labor of millions of enslaved Africans; blending (creolization) of indigenous, European, and African cultures has resulted in unique cultural expressions
- Region currently composed of 20 independent countries and several dependent territories
- The region exhibits an economic transformation, as agriculture slowly replaced by manufacturing, offshore banking, and tourism

Research or Term Paper Ideas

- Engage in library research to learn more about Maroon societies in the Caribbean. How did they organize their societies? How did they maintain their distinctive identities over the decades or centuries? How have they been incorporated in the wider society? How has globalization affected the continuance of Maroon societies?

- Compare and contrast U.S. policy toward Cuban immigrants with its policy toward Haitian immigrants.

- Engage in library research to compare the biological diversity of tropical rainforests with that of plantations. What does this suggest for our economic and environmental policies?

- Compare and contrast the establishment of free trade zones in the Caribbean with maquiladoras in Mexico. How are both economic processes tied to globalization?

Practice Quizzes

Answers appear at the end of this book

Vocabulary Matching: Match the term to its definition

A. Brain drain
B. Brain gain
C. Capital leakage
D. Chain migration
E. Creolization
F. Greater Antilles
G. Hurricanes
H. Indentured labor

I. Lesser Antilles
J. Maroons
K. Mono-crop production
L. Monroe Doctrine
M. Neocolonialism
N. Offshore banking
O. Plantation America
P. Remittances

1. _____ A pattern of migration in which one family member at a time is brought over to the new country; this can create a link between a source country (such as the Dominican Republic) and a destination (such as New York City).

2. _____ Storm systems with an abnormally low-pressure center sustaining winds of 74 mph or higher; they are especially a problem in the Caribbean and the Atlantic coast of North America.

3. _____ The blending of African and European cultures in the Caribbean.

4. _____ Escaped slaves who established communities in the Caribbean.

5. _____ The four large Caribbean islands of Cuba, Jamaica, Hispaniola, and Puerto Rico.

6. _____ The official United States policy that stated that the United States would not tolerate European military involvement in the Western Hemisphere.

7. _____ A reliance on a single commodity, such as sugar, especially in the context of plantation agriculture.

8. _____ Workers contracted to work on estates for a set period of time.

9. _____ Economic and political strategies by which powerful states indirectly (and sometimes directly) extend their influence over other weaker states.

10. _____ A cultural region located in the coastal zone of the Americas and was characterized by European-owned land worked by African laborers producing agricultural products for export.

11. _____ Monies sent by immigrants to their country of origin to support family members left behind.

12. _____ The return migration of Caribbean peoples from North America and Europe back to their homelands.

13. _____ Financial institutions in the Caribbean that offer specialized services that are confidential and tax-exempt.

14. ____ The huge gap between gross receipts and total tourist dollars that remain in the Caribbean; it is caused by the fact that profits go to the multinational tourist and hospitality corporations, leaving only the income from low-wage jobs (such as hotel maids) in the region.

15. ____ Migration of best educated people from developing countries to developed nations where economic opportunities are greater.

Multiple Choice: *Choose the word or phrase that best answers the question*

1. When Europeans began to colonize the Americas, where did they start?
    a. Latin America
    b. The Caribbean
    c. The east coast of what is now Canada
    d. The west coast of what is now the United States
    e. Yucatan Peninsula

2. In which part of the Caribbean would you find the majority of the region's population, arable lands, and large mountain ranges?
    a. Bahamas
    b. Greater Antilles
    c. Lesser Antilles
    d. The Rimland
    e. Virgin Islands

3. What feature defines seasons in the Caribbean?
    a. Temperature variation
    b. Shifting wind patterns
    c. Changes in rainfall
    d. A and B above
    e. A, B, and C above

4. Which of the following statements about hurricanes in the Caribbean is/are true?
    a. Caribbean hurricanes originate off the west coast of Africa
    b. Hurricanes typically enter the Caribbean through the Lesser Antilles
    c. The hurricane season lasts from July to October
    d. A and B above
    e. A, B, and C above

5. Today, where are the tropical forests most commonly found in the Caribbean?
    a. Bahamas
    b. Greater Antilles
    c. Lesser Antilles
    d. The Rimland
    e. Virgin Islands

6.   What is the most significant demographic trend in the Caribbean?
     a.   Decline in fertility
     b.   Drop in population
     c.   Increase in family size (TFR) throughout the region
     d.   Increased immigration from Russia since the breakup of the Soviet Union
     e.   Increasing rate of natural increase (RNI)

7.   Which of the following matchups between Caribbean migrants and their destinations is/are true?
     a.   Barbados residents go to England; Surinamese go to the Netherlands
     b.   Cubans go to Russia
     c.   French Guianans go to France
     d.   A and B above
     e.   A and C above

8.   Caribbean cities in areas colonized by the Spanish tend to look like those in which other world region?
     a.   Latin America
     b.   Sub-Saharan Africa
     c.   South Asia
     d.   North America
     e.   Europe

9.   In which country of the Caribbean are squatter settlements LEAST common?
     a.   Bahamas
     b.   Cuba
     c.   Dominican Republic
     d.   Haiti
     e.   Jamaica

10. Why were indentured workers from South Asia needed in the Caribbean?
     a.   Illness killed most of the slaves and peasant farmers
     b.   Slaves had been freed, and slavery made illegal, eliminating that labor source
     c.   The colonizers found additional markets and needed more workers for their expanded production
     d.   The work became harder as lands became less productive
     e.   All of the above

11. Where in the Caribbean did Maroons tend to settle?
     a.   Along the rivers?
     b.   In the cities
     c.   On the best farmlands
     d.   On the coasts
     e.   In isolated areas

12. Why were plantations described as "monocrop?"
     a.   Because of the number of markets for their products
     b.   Because of their ranking in terms of agricultural productivity, compared to other types of agricultural
     c.   Because the most common number of commodities grown there was 1 (one)
     d.   Because they typically had just one manager
     e.   Because they were the number one favorite type of agriculture in the region

13.   Which of these are grown in the Caribbean?
    a.   Bananas
    b.   Coffee
    c.   Sugar
    d.   A and B above
    e.   A, B, and C above

14.   What makes Cuba distinctive among Caribbean countries?
    a.   It has the highest elevation
    b.   It has the highest percentage of rainforests remaining
    c.   It has the smallest population in the region
    d.   It is communist
    e.   It is the only monarchy in the region

15.   Which of the following is not a current economic activity in the Caribbean?
    a.   Computer software development
    b.   Banking
    c.   Assembly-plant industrialization
    d.   Sugarcane and coffee cultivation
    e.   Tourism

Summary of Caribbean

Total Population of Caribbean: 43 million

Population Indicators for Caribbean

	Highest (country and value)	Region Average	Lowest (country and value)
Population 2010 (millions)	Cuba: 11.2	1.66	Montserrat: 0.005
Density per sq km	Bermuda: 1264	241.31	Suriname; French Guiana: 3
RNI	French Guiana: 2.4	1.03	Cuba: 0.3
TFR	Montserrat: 1.2	2.17	French Guiana: 3.6
Percent Urban	Anguilla; Bermuda; Cayman; Guadeloupe: 100%	64%	Trinidad and Tobago: 12%
Percent < 15	Haiti; Belize: 37%	26%	Bermuda: 18%
Percent > 65	Bermuda: 15%	0.04%	French Guiana; Haiti; Turks and Caicos: 4%
Net Migration (per 1000;2000-05)	Cayman: 16	-2.37	Guyana: -10.5

Development Indicators for Caribbean

	Highest (country and value)	Region Average	Lowest (country and value)
GNI per capita/PPP 2008	Bermuda: $69,900	$15,943	Haiti: $1,300
GDP Avg. Annual Growth (2000-2008)	Trinidad and Tobago: 6.6%	4.03%	Haiti: 0.5%
Human Development Index (2007)	Barbados: 0.903	0.800	Haiti: 0.532
Percent Living below $2/day	Haiti: 72.1%	26.9%	Jamaica: 5.8%
Life Expectancy 2010	Anguilla; Bermuda; Cayman: 81	74	Haiti: 61
< 5 Mortality 2008	Haiti: 72 per 1000	44 per 1000	Cuba: 6 per 1000
Gender Equity 2008	Suriname: 114	100	Cayman: 90

Chapter 6

Sub-Saharan Africa

Learning Objectives

➢ This chapter introduces Africa south of the Sahara.
➢ The trade of enslaved Africans is an important aspect of the region's history, an aspect that continues to have significance.
➢ European colonialism, and its legacy, has significantly influenced the political and economic development of many countries in the region.
➢ The region provides valuable lessons for our understanding of health and medical geographies.
➢ Key topics to be developed include desertification, agricultural density, shifting cultivation (swidden), and structural adjustment programs.

Chapter Outline

1. **Introduction:** sub-Saharan Africa is composed of 48 states and one territory; it has a population of nearly 865 million; compared to Latin America and the Caribbean, it is relatively poor and less urbanized
    A. Sub-Saharan Africa—the portion of the African continent south of the Sahara desert—is a commonly accepted world region; its unity is based on similar livelihoods and a shared colonial heritage; however, no common religion, language, philosophy, or political system ever united the region
        a. North Africa is sometimes considered along with sub-Saharan Africa; however, because of multiple reasons, North Africa has stronger regional associations with Southwest Asia
    B. Culturally, the African presence is well-represented throughout the world; music and the arts are strongly global; economically, however, the economies of Africa remain marginal at the global level
2. **Environmental Geography:** Africa is the largest landmass straddling the equator; it is often called the *plateau continent*
    A. **Plateaus and Basins:** Africa's physical geography is dominated by plateaus and elevated basins; elevations generally increased toward the south and east
        a. **The Great Escarpment:** rims southern Africa, beginning in southwest Angola and ending in northeast South Africa; it has been an impediment to coastal settlement
    B. **Watersheds**: sub-Saharan Africa lacks the broad, alluvial lowlands that influence settlement patterns in other regions; there are four major river systems—the Congo, Nile, Niger, and Zambezi; other, small rivers include the Orange, the Senegal, and the Limpopo
        a. **The Congo (Zaire) River:** this is the largest watershed in the region in terms of drainage and volume of flow; it is second only to the Amazon in the world; rapids and falls limit navigability
        b. **The Nile River:** the world's longest river; important source of water for Egypt and Sudan
        c. **The Niger:** critical source of water for Mali and Niger; originates in the Guinean Highlands
        d. **The Zambezi River:** a major supplier of commercial energy; numerous hydroelectric dams are located along its course; the Victoria Falls are located in the Zambezi
    C. **Soils:** soils are relatively infertile, in part because materials and nutrients have been leached away; some areas, particularly around the Rift Valley, have better soils because of volcanic activity
    D. **Climate and Vegetation:** sub-Saharan Africa lies in the tropical latitudes; generally warm year round; consistent rainfall along the equator, becoming more seasonal to the north and south; arid,

desert conditions exist toward northern and southern edge of continent; mountain zones exhibit altitudinal zonation

- a. **Tropical Forests:** the world's second-largest expanse of humid equatorial rain forest, the Ituri, lies in the Congo Basin; commercial logging and agricultural clearing have degraded the western and southern fringes of the forest; much of the northeastern section remains intact; major national parks have been established in the region; poor infrastructure and political chaos has made large-scale logging impossible; conflict has also made conservation difficult
- b. **Savannas:** surround Central African rain forest belt; wetter savannas next to forests, fewer trees and shorter grasses farther north and south
- c. **Deserts:** tropical Africa is bracketed by several deserts
  - i. **The Sahara:** the world's largest desert, spans the landmass from Atlantic coast to Red Sea and around the Horn of Africa
  - ii. **Namib Desert:** located along coast of Namibia; temperatures generally mild
  - iii. **Kalahari:** located east of the Namib Desert; technically not a desert because it receives slightly more than 10 inches of rainfall annually
- E. **Africa's Environmental Issues:** because much of the region's population earns its livelihood from the land, environmental changes are very significant
  - a. **Desertification:** the expansion of desert-like conditions as a result of human-induced degradation; deforestation is major cause, also caused by improper cultivation, cultivation of non-native species, and over-grazing
  - b. **Deforestation:** unlike tropical America, forest clearance in the wet and dry savanna is of greater local concern than the limited commercial logging of the rain forest; deforestation in the savanna woodlands aggravates problems of increased runoff, soil erosion, and shortages of biofuels (wood and charcoal used for household energy needs); in some regions, community-based non-governmental organizations have mobilized to re-plant areas; deforestation is significant in the southern fringes of Central Africa's Ituri forest; two smaller rain forests, one along the Atlantic coast from Sierra Leone to western Ghana and the other along the eastern coast of Madagascar, have nearly disappeared as a result of commercial logging and clearance for agriculture
  - c. **Wildlife Conservation:** sub-Sahara has significant biodiversity and abundance of animals, especially of large mammals; many are under threat due to habitat loss and poaching; a number of wildlife preserves have been developed, mostly in southern Africa; these are popular tourist destinations
- F. **Global Warming and Vulnerability:** This region is the lowest emitter of greenhouse gases but is likely to experience greater human vulnerability to global warming; this is because of the region's limited resources to respond and adapt to climate change; some regions might receive more rainfall, but many others might receive less, thus limiting productivity; increases in famine and disease are possible
3. **Population and Settlement:** the region is experiencing rapid population growth; has a young population (43 percent of people under 15 years of age); large family size (approximately five or six children); child and mortality rates are high; life expectancy is low (52 years); relatively low population density (similar to the United States); physiological density—the number of people per unit of arable land, is much higher; its agricultural density—the number of farmers per unit of arable land, is also high
   - A. **Population Trends and Demographic Debates:** a main question: is sub-Saharan Africa overpopulated?
     - a. **Family size:** large families have been preferred because of cultural practices, rural lifestyles, high child mortality, and economic realities
     - b. **Impact of AIDs on Africa:** as of 2008, two-thirds of HIV/AIDs cases in the world were found in this region; infection rates high enough to slow population growth in some locations;

advanced medicines are too costly for most people; education has helped slow the spread of the disease

- B. **Patterns of Settlement and Land Use:** Most Africans live in rural settlements; population concentrations are highest in West Africa, highland East Africa, and the eastern half of South Africa; population concentrations are largely the result of better soils and agriculture
- C. **Agricultural Subsistence:** over much of continent, African agriculture remains relatively unproductive; rural population densities tend to be low; in some areas, *swidden* (shifting cultivation) is practiced; women are usually subsistence farmers
- D. **Plantation Agriculture:** remains critical to the economies of many states; especially important since region has few competitive industries; many countries rely on one or two export crops
- E. **Pastoralism:** animal husbandry (care of livestock) is extremely important for the region, especially in semi-arid regions; however, large expanses of sub-Saharan Africa have been off-limits to cattle because of tsetse flies, which spread sleeping sickness to cattle, humans, and some wildlife
- F. **Urban Life:** sub-Saharan Africa is the least urbanized region in the developing world; urban areas are increasing; rural-to-urban migration, industrialization, and refugee flows account for much growth; cities are becoming over-urbanized because of unplanned, rapid growth
  - a. **West African Urban Traditions:** West Africa had urban tradition long before European colonialists; many cities combine Islamic, European, and national elements
  - b. **Urban Industrial South Africa:** most cities in southern Africa are colonial in origin; South Africa is one of most urbanized states in the region—its urban economy is based largely on rich mineral resources; the form and layout of South African cities continues to reflect legacy of *apartheid*, an official policy of racial segregation imposed by Europeans

4. **Cultural Coherence and Diversity:** No overall cultural unity to sub-Saharan Africa; most African kingdoms and empires of past were limited to distinct sub-regions; traditional African religions also largely limited to local areas; some trade languages have dispersed over wider areas, but no indigenous language spans entire continent
- A. **Language Patterns:** complex pattern with local language groups, African trade languages, and languages introduced from Europe and Asia
  - a. African Language Groups: three of six language groups are unique to region (Niger-Congo, Nilo-Saharan, and Khoisan), while three others (Afro-Asiatic, Austronesian, and Indo-European) are more closely associated with other world regions
  - b. **Language and Identity:** ethnic identity and linguistic affiliation have historically been fluid; distinct tribes consisted of groups of families or clans with common kinship, language, and definable territory; European colonial administrators began process of establishing fixed social orders as means of control; many tribes were artificially divided, meaningless names applied, and territorial boundaries misinterpreted; social boundaries have become more stable now
  - c. **European Languages:** colonial powers promoted use of European languages, especially in government and education; after decolonization, many countries in region continued to use these languages
- B. **Religion:** indigenous African religions generally classified as *animist*, a misleading catchall term used to classify all local faiths that do not fit into "world religions"; most animist religions centered on worship of nature and ancestral spirits
  - a. **Introduction and Spread of Christianity:** arrived around AD 300 in Ethiopia and central Sudan; European settlers and missionaries introduced Christianity to other parts beginning in 1600s; today, Christianity is spread irregularly across the non-Islamic portion of the region
  - b. **Introduction and Spread of Islam:** began to diffuse around 1,000 years ago; diffused along trade routes; later kingdoms and empires converted to Islam; today, orthodox Islam prevails throughout much of the Sahel; further south Muslims are mixed with Christians and animists; numbers are growing

    c. **Interaction Between Religious Traditions:** religious conflict most acute in northeastern Africa, between Muslims and Christians; elsewhere a pattern of peaceful coexistence is found; many syncretic forms of religious expression are emerging

  C. **Globalization and African Culture:** initial diffusion of African culture the result of the slave trade; African diaspora reflects blending of African cultures with Amerindian and European; African culture within Africa also influenced by foreign elements, including European languages, religions, and dress

    a. **Popular Culture in Africa:** dynamic mixture of global and local influences, found in music and dance, for example

    b. **Music in West Africa:** Nigeria is musical center of West Africa, with well-developed and cosmopolitan recording industry; contemporary African music is both commercially and politically important

    c. **The Pride of Runners:** Ethiopia and Kenya have produced many of the world's greatest distance runners; in these countries, running is a national pastime

5. **Geopolitical Framework:** over the millennia, many diverse ethnic groups formed in the region; with arrival of Europeans, patterns of social and political relations changed

  A. **Indigenous Kingdoms and European Encounters:** Nubia was the first significant state to emerge in sub-Saharan Africa, located in central and northern Sudan around 3,000 years ago; other states located throughout West Africa; most associated with trans-Saharan trade routes; some African states participated in slave trade and increased military and economic power

    a. **Early European Encounters:** it took Europeans many centuries to gain effective control of the region

    b. **Disease Factor:** malaria and other tropical diseases killed many Europeans, thus slowing colonization; pursuit of colonies led to development of quinine to protect against malaria

  B. **European Colonization:** Occurred rapidly after 1880s, termed the "scramble for Africa"; European colonization of Africa associated with other global events, such as formation of new European countries seeking colonies

    a. **Berlin Conference:** 13 countries convened in Berlin in 1884 to established rules for colonial division of Africa

    b. **Establishment of South Africa:** South Africa was one of oldest colonies in Africa, and first to obtain political independence in 1910; in 1948, Afrikaners' National Party established apartheid (racial segregation) and established black *homelands* for black Africans

  C. **Decolonization and Independence:** began quickly and relatively peacefully in 1957; many independence movements associated with broader *Pan-African Movement*; in 1963, the Organization of African Unity (OAU) was formed, later renamed the *African Union* (in 2002); regional economic organizations have been established, including the Economic Community of West African States (ECOWAS), Southern African Development Community (SADC), Economic Community of Central African States (ECCAS), and the East African Community (EAC)

  D. **Southern Africa's Independence Battles:** Independence was heavily contested and violent in southern Africa, especially Southern Rhodesia (modern-day Zimbabwe) and former Portuguese colonies of Angola and Mozambique

  E. **Apartheid's Demise in South Africa:** opposition to apartheid began in 1960s with internal pressure from blacks, coloureds, and Asians; Nelson Mandela—a key black South African anti-apartheid leader—was freed in 1990 after 27 years of political imprisonment; international pressure also a factor; in 1994, free elections led to Mandela assuming presidency; apartheid eliminated, but legacy remains

  F. **Enduring Political Conflict:** Although independence for most states was relatively peaceful, post-colonial decades have been more unstable

    a. **Tyranny of the Map:** most states reflect legacy of colonialism and imposition of political boundaries; in 1963, the OAU agreed to maintain these borders; tribalism, or loyalty to ethnic group rather than state, has emerged as problem; conflicts have led to numerous refugees (people who flee their state because of well-founded fear of persecution based on race, ethnicity, religion, or political orientation) and internally displaced persons (people who have fled from conflict but remain in their state)

    b. **Ethnic Conflicts:** more than half of the states in the region have experienced wars or serious insurrections since 1995; peace has returned to some countries; conflicts partially related to resources, but relationship is complex; ethnic identity also a contributing factor, but equally complex

    c. **Secessionist Movements:** problematic African political boundaries, established by European colonialists and agreed upon by OAU following independence, continue to led to tensions; Eritrea seceded from Ethiopia, the only territory to achieve this goal

6. **Economic and Social Development:** sub-Saharan Africa is poorest and least-developed world region; many states have experienced declines in life expectancy since 1990 and income levels remain low; in 1990s, the International Monetary Fund and the World Bank promoted a series of *structural adjustment programs*; these are designed to reduce government spending, cut food subsidies, and encourage private sector initiatives; these policies have led to hardships at the local and family level; some signs of economic growth in some countries

  A. **Roots of African Poverty:** traditional explanations focused on environmental factors, but these can be overcome; other factors include legacy of slave trade, colonization (and especially plantations and mining activities), and lack of infrastructure development

    a. **Failed Development Policies:** policies designed to promote economic growth and to provide cheap supply of stable foods in urban areas largely failed; governments kept prices artificially low, thus hurting rural farmers (who compose majority of population); local currencies kept at artificially elevated levels, which benefited the elite but undercut exports

    b. **Millennium Development Goals:** new approach adopted—the Millennium Development Goals, as global effort to foster development; established as part of United Nations effort to reduce extreme poverty; unlikely that states in region will meet goals; international interest in Africa's overall development is growing, with aid being directed toward disease control, health care, and basic education

    c. **Corruption:** particularly rampant in several African countries; some refer to *kleptocracy*, a state in which corruption is so institutionalized that politicians and government bureaucrats siphon off a huge percentage of country's wealth

  B. **Links to World Economy:** trade connections with the world are limited; account for less than 2 percent of global trade; measures of connectivity for region are low; Africa lacks basic infrastructure to facilitate trade

    a. **Aid Versus Investment:** Africa is linked globally more by aid than by flow of goods or foreign investment; poverty and political instability discourage investment

    b. **Debt Relief:** development strategy proposed to reduce debt levels; interest payments are very high

  C. **Economic Differentiation Within Africa:** considerable differences in economic and social development persist; small island states (e.g., Mauritius and the Seychelles) are economically better off; millions of Africans live in extreme poverty; some places are experiencing social and economic gains, including South Africa, which has a well-developed and well-balanced industrial economy

    a. **Oil and Mineral Producers:** a handful of countries benefit from substantial oil reserves, including Gabon, the Republic of the Congo, and Equatorial Guinea; both Namibia and Botswana have abundant mineral reserves; all of these countries benefit also from smaller population sizes

    b. **Leaders of ECOWAS:** Nigeria, the most populous country, is a core member of ECOWAS; the state has the largest oil reserves and is an OPEC member; other important commercial states include the Ivory Coast and Ghana

    c. **East Africa:** Kenya historically has been a major economic leader of the region, but has experienced economic decline and political tension since the 1990s; Uganda and Tanzania are improving

D. **Measuring Social Development:** by world standards, measures of social development are low; some positive trends, with regard to child survival, education, and gender equity, are apparent

    a. **Life Expectancy and Health Issues:** life expectancy for sub-Saharan Africa is lowest for any world region; factors contributing include extreme poverty, environmental hazards (drought), and various environmental and infectious diseases; problems exacerbated by lack of adequate and accessible health care facilities

    b. **Meeting Educational Needs**: providing universal education is problematic for youthful population; gender equity, especially in West Africa, has improved

    c. **Women and Development:** women are often invisible contributors to local and national economies; social position of women is complex—women traders have considerable political and economic power and female labor force participation is relatively equal; however, prevalence of polygamy, practice of "bride price," and denial of property inheritance continues, as does discrimination; most controversial issue is female circumcision practiced throughout Sudan, Ethiopia, Somalia, Eritrea, and parts of West Africa

        i. **Building from Within:** Support groups and networks have development, raising women's consciousness, offering women micro-credit loans for small businesses, and harnessing economic power

Summary

- Africa is the largest landmass straddling the equator; it is called the plateau continent; key environmental issues include desertification, deforestation, and drought
- With 865 million people, sub-Saharan Africa is fastest growing region
- Sub-Saharan Africa is culturally diverse
- Since 1995, numerous wars and conflicts have afflicted the region
- The region's connections to the global economy are weak; limited foreign direct investment, especially from China, is increasing
- Poverty is the region's most pressing issue

Research or Term Paper Ideas

- Learn more about wildlife and biodiversity within Africa. Choose a particular ecosystem, such as the tropical rain forest or savanna. Identify the habitat (including, for example, meteorological and climatic factors, and vegetation). What types of animals are found in the ecosystem? What are the current threats to both the ecosystem and the wildlife?

- Conduct research on female circumcision in Africa. What is its origin? What does it entail? How common was, and is, the practice? How does this practice reflect the broader issue of gender roles and expectations in Africa? What is being done to eliminate this practice?

- Obtain data on current incidences of HIV/AIDs in Sub-Saharan Africa. What population groups are most impacted? What measures are currently in place to help prevent the spread of the disease?

Practice Quizzes

Answers appear at the end of this book

Vocabulary Matching: Match the term to its definition

A. Agricultural density
B. Apartheid
C. Berlin Conference
D. Desertification
E. Genocide
F   Kleptocracy
G. Millennium Development Goals
H. Pan-African Movement

I. Pastoralists
J. Physiological density
K. Refugees
L. Sahel
M. Structural adjustment programs
N. Swidden
O. Transhumance
P. Tribalism

1. _____   The expansion of desert-like conditions as a result of human-induced degradation.

2. _____   A zone of ecological transition between the Sahara to the north and the wetter savannas and forests to the south.

3. _____   The seasonal movement of animals between wet- and dry-season pastures.

4. _____   People who flee their country because of a well-founded fear of persecution based on race, ethnicity, religion, or political affiliation, or because of war.

5. _____   A population statistic that relates the number of people in a country to the amount of arable land.

6. _____   Loyalty to an ethnic group rather than to the state.

7. _____   The number of farmers per unit of arable land.

8. _____   Farmers who raise large animals (especially cattle, camels, sheep, and goats) for their sustenance and livelihood.

9. _____   The policy of racial separation or segregation in South Africa that was in place for nearly 50 years.

10. _____   An agricultural practice that involves burning the natural vegetation to release fertilizing ash and then planting indigenous crops (maize, beans, yams, etc.); once the nutrients on a plot of land have been exhausted, it is temporarily abandoned until its fertility returns.

11. _____   A state where corruption is so institutionalized that large percentages of the country's wealth is siphoned off by politicians and government bureaucrats.

12. _____   The 1884 meeting during which Africa was divided into European colonial territories; these boundaries ignored cultural affiliations, and many of Africa's civil conflicts are linked to this boundary-making.

13. _____ A program of the United Nations aimed at fostering development and reducing extreme poverty by 2015 in Sub-Saharan Africa.

14. _____ The deliberate and systematic killing of a racial, political, or cultural group.

15. _____ Controversial economic measures designed to reduce government spending and encourage private sector initiatives, and refinance foreign debt; typically, these IMF and World Bank policies trigger drastic cutbacks in government-supported services and food subsidies, which disproportionately harm the poor.

Multiple Choice: *Choose the word or phrase that best answers the question*

1. What global landmark passes through Sub-Saharan Africa?
   a. Equator
   b. International dateline
   c. Arctic Circle
   d. A and B above
   e. A and C above

2. What makes Sub-Saharan Africa important from an anthropological perspective?
   a. Domestication of plants and animals occurred in this region first
   b. Human origins are traceable to this region
   c. It has the largest population of any region in the world
   d. People in this region speak the highest number of languages of any world region
   e. This is the source region of Christianity

3. What is the major environmental issue in Sub-Saharan Africa, and where is it most serious?
   a. Desertification in the Sahel
   b. Drought in the Congo
   c. Flooding in South Africa
   d. Hurricanes on the Horn of Africa
   e. Water pollution in North Africa

4. Deforestation in Sub-Saharan Africa causes which of the following problems?
   a. Moisture loss
   b. Flooding in South Africa
   c. Soil erosion
   d. A and B above
   e. A, B, and C above

5. What is the longest river in Africa and the world?
   a. Congo
   b. Niger
   c. Nile
   d. Senegal
   e. Zambezi

6. What is Sub-Saharan Africa's (and the world's) largest desert?
   a. Congo
   b. Angolan
   c. Kalahari
   d. Namib
   e. Sahara

7. Why is Sub-Saharan Africa experiencing population growth?
   a. High birth rate coupled with low death rate
   b. High rate of immigration
   c. High rate of urbanization
   d. Preference for large families
   e. All of the above

8. How do the majority of the people of Sub-Saharan Africa earn a living?
   a. Mining
   b. Manufacturing
   c. Forestry
   d. Fishing
   e. Agriculture

9. In which part of Africa has the impact of HIV/AIDS been most devastating?
   a. North
   b. South
   c. The Center
   d. The Horn
   e. The West

10. What is Apartheid?
    a. A political subdivision in Tanzania
    b. A political party in Ethiopia
    c. A policy of racial separation in South Africa
    d. A landform type found on the west coast of Africa
    e. The Swahili name for bird flu

11. More than anything else, which of the following has linked Africa to the Americas and Europe?
    a. Cocoa
    b. Foreign aid
    c. Language
    d. Rubber
    e. Slavery

12. What African language became the most widely spoken sub-Saharan language in the region?
    a. Mandingo
    b. Yoruba
    c. Amharic
    d. Swahili
    e. Igbo

13.   What is the significance of Nubia?
    a.   It was the only country of sub-Saharan Africa to escape colonization
    b.   It was the first significant state to emerge in sub-Saharan Africa
    c.   It was the most populous country of Africa
    d.   It is the leader of the African Union
    e.   It is the largest country in sub-Saharan Africa

14.   What happened at the Berlin Conference?
    a.   The African Union was founded
    b.   African countries agreed to a guest-worker agreement with the European Union
    c.   European colonial powers divided Africa among themselves
    d.   African and European countries drafted a peace treaty to end the conflict between the two regions
    e.   South Africa decided to end Apartheid

15.   Which of the following have been identified as reasons for poverty in sub-Saharan Africa?
    a.   Unfavorable environmental conditions
    b.   Slavery
    c.   Failed development policies and corruption
    d.   A and B
    e.   A, B, and C above

Summary of Sub-Saharan Africa

Total Population of Sub-Saharan Africa: 865 million

Population Indicators for Sub-Saharan Africa

	Highest (country and value)	Region Average	Lowest (country and value)
Population 2010 (millions)	Nigeria: 158.3	17.64	São Tomé and Principe: 0.2
Density per sq km	Mauritius: 628	93	Botswana; Mauritainia; Namibia: 3
RNI	Niger: 3.5	2.38	Seychelles: 1.0
TFR	Niger: 7.4	4.8	Mauritius: 1.5
Percent Urban	Reunion: 92%	40%	Burundi: 10%
Percent < 15	Niger; Uganda: 49%	41%	Mauritius: 22%
Percent > 65	Seychelles: 10%	3%	Angola; Eritrea; Ivory Coast; Niger; Rwanda; Senegal; Sierra Leone: 2%
Net Migration (per 1000;2000-05)	Liberia: 13.3	-0.52	Zimbabwe: -11.1

Development Indicators for Sub-Saharan Africa

	Highest (country and value)	Region Average	Lowest (country and value)
GNI per capita/PPP 2008	Equatorial Guinea: $21,700	$3,332	Dem. Rep. of Congo: $280
GDP Avg. Annual Growth (2000-2008)	Angola: 13.5%	4.59%	Zimbabwe: -5.7%
Human Development Index (2007)	Seychelles: 0.845	0.523	Niger: 0.340
Percent Living below $2/day	Tanzania: 96.6%	69.8%	Seychelles: <2%
Life Expectancy 2010	Reunion: 78	55	Lesotho: 41
< 5 Mortality 2008	Angola: 22 per 1000	126 per 1000	Mauritius: 17 per 1000
Gender Equity 2008	Lesotho: 105	90	Chad: 64

Chapter 7

Southwest Asia and North Africa

Learning Objectives

➤ This chapter introduces Southwest Asia and North Africa, a region commonly known as the Middle East

➤ You will be introduced to the complex ways in which this region has been affected by globalization, and itself impacts globalization

➤ You will gain an understanding of this region's significant role in both world trade and world religions

➤ This chapter provides insight into the many political conflicts that affect the region, and by extension, how these conflicts affect the world more broadly

➤ The following terms and concepts will be introduced and emphasized: Islamic fundamentalism, the Maghreb and Levant, salinization, pastoral nomadism, and monotheism

Chapter Outline

1. **Introduction**
   A. The region extends for 4,000 miles, stretching from Morocco's Atlantic coast through to Iran's eastern border with Pakistan; regional boundaries and terminology defy easy definition
      a. The region is often termed the *Middle East*; this term carries a Eurocentric component and excludes North Africa and the states of Turkey and Iran; the text uses the term "Southwest Asia and North Africa"
      b. Culturally, diverse languages, religions, and ethnic identities have been important within the region; the region is considered a key global *culture hearth*
      c. There are numerous zones of conflict in the region; these are in part related to the rise of Islamic fundamentalism
      d. The region is significant, globally, because of its strategic location and oil and natural gas reserves
      e. The region's environment provides challenges; availability of water is a significant concern

2. **Environmental Geography:** the regional terrain varies greatly, with rocky plateaus and mountain ranges, but also deserts; the climate varies also; a dominant theme is that the legacy of human settlement has left its mark on a fragile environment and ecological problems lie ahead
   A. **Legacies of a Vulnerable Landscape:** lengthy human settlement in a marginal land has resulted in deforestation, soil salinization and erosion, and depleted water resources
      a. **Deforestation and Overgrazing:** this is an ancient problem in the region; growing demands for agricultural land causes upland forests to be removed and replaced with fields; some forests remain in mountainous regions; overgrazing of livestock contributed to forest loss
      b. **Salinization:** the buildup of toxic salts in the soil; a historical problem in the region because of irrigation; problem particularly acute in Iraq
      c. **Managing Water:** refers to practices of modifying drainage systems and water flows; *qanat system*—tapping into groundwater through series of sloping tunnels; *fossil water*—water supplies stored underground during earlier wetter periods, being tapped and transferred; *hydropolitics*—the interplay of water resource issues and policies, particularly important between countries that share watersheds
   B. **Regional Landforms:** a diversity of environmental settings and landforms
      a. **Maghreb:** (Arabic for "western island") includes Morocco, Algeria, and Tunisia; dominated by Atlas Mountains; Levant (mountainous eastern Mediterranean region; Arabian

Peninsula—eastward sloping plateau; highlands in Oman and Yemen; Iranian and Anatolian (sometimes called Asia Minor) Plateaus—both geologically active and prone to earthquakes

C. **Patterns of Climate:** aridity dominates much of the region; driest conditions across North Africa (especially the Sahara); Mediterranean climate along the Levant coastline and portions of northern Syria, Turkey, and northwestern Iran

D. **Global Warming in Southwest Asia and North Africa:** projected changes in global climate will tend to aggravate already-existing environmental problems; temperature changes thought to be more significant than changes in precipitation—higher temperatures will exacerbate evaporation rates and stress crops, grasslands, and other vegetation; hydroelectric potential will be reduced; sea-level changes pose particular problem to Nile Delta region; might overall contribute to political tensions

3. **Population and Settlement:** population density within region varies with availability of water

A. Regional population of around 450 million; great variation in distribution and density; overall population density relatively modest, but physiological density ranks among highest on Earth

   a. Urban population highly uneven; less than two-thirds of population is urbanized; urban crowding significant problem

   b. Two dominant clusters of settlement, both based on availability of water: (1) coastal region of Maghreb and (2) Nile River Valley; other smaller concentrations through Tigris and Euphrates valley, Yemen Highlands, near oases

B. **Water and Life:** water and life closely linked across rural settlement patterns

   a. Region is home to one of world's earliest hearths of *domestication* (purposive selection and breeding of plants and animals); around 10,000 years ago, domestication began in an area known as the *Fertile Crescent*—an ecologically diverse zone that stretches from the Levant through northern Syria and into Iraq; practices spread to nearby valleys, including Tigris and Euphrates (Mesopotamia) and Nile Valley

   b. **Pastoral Nomadism:** traditional form of highly mobile and flexible subsistence agriculture in which practitioners depend on seasonal movement of livestock; lifestyle and practice in decline—still prevalent among Arabian Bedouins, North African Berbers, and Iranian Bakhtiaris; declining because of reduced demand for beasts of burden, competing land uses, overgrazing, and constricting political borders

   c. **Transhumance:** seasonal movement of animals between wet-season and dry-season pastures—practiced especially in mountainous regions such as Atlas Mountains and Anatolian Plateau

   d. **Oasis Life:** permanent oasis settlements scattered throughout arid region; high groundwater levels or modern deep-water wells provide reliable moisture; settlement dominated by close-knit families—mostly subsistence but some commercial trade

   e. **Exotic Rivers** (rivers that originate in distant, moisture regions); examples include Tigris, Euphrates, and Nile; existence makes settlement possible, but vulnerable to overuse, particularly if irrigation results in salinization

C. **Challenge of Dryland Agriculture:** these regions depend on seasonal moisture; include better-watered valleys and coastal lowlands of northern Maghreb, the Levant, and Anatolian and Iranian Plateau's uplands; variety of crops and livestock—vulnerable to drought; mechanization and crop specialization transforming agricultural practices

D. **Many-Layered Landscapes:** The Urban Imprint

   a. **Long Urban Legacy:** cities began in Mesopotamia (Iraq) around 3500 BC, in Egypt by 3000 BC; early cities were centers of politics and religion; trade centers along caravan routes; religious study (especially Islam); European colonialism influenced later architectural forms

   b. **Signatures of Globalization:** Urban centers are focal points of economic growth, drawing in rural population; many cities, such as Cairo, Algiers, and Istanbul, experience over-urbanization problems; oil rich states of Persian Gulf display most significant changes—new infrastructure, modern urban designs and futuristic architectural styles incorporated in cities such as Abu Dhabi, Doha, Kuwait City, and Riyadh

E. **Recent Migration Patterns:** new patterns of migration: (1) rural-to-urban shift; (2) in-migration of laborers, especially from South and Southeast Asia moving to oil-rich countries; (3) out-migration of residents moving to other parts of the world, especially of Turkish guest workers to Europe; (4) wealthier residents leaving because of political instability—many destined to Europe or North America; (5) refugee flows and internally displaced persons resultant from wars in Iraq and Afghanistan.

F. **Shifting Demographic Patterns:** high population growth remains an issue, but exhibits regional variation; some regions, such as Iran, Turkey, and Tunisia, experience declines in birth rates; other areas, including Yemen, retain high growth rates; some countries face problem of providing food, jobs, and housing for young populations (e.g., Egypt)

4. **Cultural Coherence and Diversity:** Although heart of Islamic and Arab worlds, reveals high degree of cultural diversity

   A. **Patterns of Religion**
      a. **Hearth of Judeo-Christian Tradition**—both Judaism and Christianity trace roots to eastern Mediterranean; neither group is numerically dominate, but both play key cultural roles; both groups practice monotheism (belief in one God)
      b. **Emergence of Islam**—Islam originated in Southwest Asia in AD 622 CE as a continuation of Judeo-Christian tradition; Muslims follow the *Quran* (or Koran), a book of revelations; Muslims practice five pillars of faith; some Islamic fundamentalists argue for theocratic state, in which religious leaders (ayatollahs) guide policy—modern day Iran is an example
         a. **Two main branches of Islam:** Shiites and Sunnis
         b. **Diffusion of Islam:** diffused widely, initially following caravan routes, military campaigns, and other trade routes; some areas declined (Iberian Peninsula), while others grew (Ottoman Empire)
      c. **Modern Religious Diversity:** Muslims form majority of population in all countries except Israel (where Judaism is dominant); divisions within Islam have created key cultural differences—Sunni Islam is numerically dominant (73 percent), compared to Shiite (23 percent); Shiite Muslims dominate in Iran, southern Iraq, Lebanon, Sudan, and Bahrain

   B. **Geographies of Language:** although often referred to as "Arab World," it is actually a linguistically complex region; individual states also reflect diversity—more than 70 languages recognized in Iran
      a. **Semites and Berbers:** Arabic-speaking Semitic peoples found from Morocco to Saudi Arabia, spread with diffusion of Islam; Hebrew, also Semitic language, is official language of Israel; older Afro-Asiatic languages endure in Atlas Mountains and parts of Sahara—these are known collectively as Berber; Berber languages are related but not mutually intelligible; most have never been written
      b. **Persians and Kurds:** Much of Iranian Plateau and nearby mountains dominated by Indo-European languages; principle language is Persian (which has been enriched by Arabic words); Persian has many distinct dialects, including *Farsi*, which is Iran's official language; Kurds of northern Iraq, northern Iran, and eastern Turkey speak Kurdish, also part of the Indo-European language family; Kurdish is shared by 10–15 million people—"Kurdistan" is often called the world's largest nation without its own political state
      c. **The Turkish Imprint:** Turkish languages are part of Altaic language family; dispersed throughout Turkey and northern Iran; other related Altaic languages include Azeri, Uzbek, and Uighur, and are found throughout parts of Southwest and Central Asia

   C. **Regional Cultures in Global Context:** cultural connections tie the region with the world; religion, especially Islam, unites people with other Muslims around the world; colonialism and recent economic and cultural ties also significant
      a. **Islamic Internationalism:** Islamic communities are established throughout the world and it is a global religion; however, Islam remains centered in Southwest Asia and North Africa
      b. **Globalization and Cultural Change:** region is struggling with growing role in global economy; technology also contributing to cultural change; hybrid forms of popular culture

are appearing (e.g., Arabic hip-hop); but conservative cultural influences are also gaining strength in some areas

5. **Geopolitical Framework:** geopolitical tensions remain high; ethnic, religious, and linguistic struggles apparent; complex ties to European colonialism remain significant source of tension, in part because of imposed political boundaries; American political power and influence is source of tension; geographies of wealth and power within region also important

    A. **Colonial Legacy:** from 1550 to 1850, much of region dominated by Turkish Ottoman Empire; from 1850 through 1950s, European colonial presence

    B. **Imposing European Power:** French colonial ties, especially in North Africa, from 1800, with France establishing *protectorates* under broader sphere of influence; Great Britain's influence in late 19ᵗʰ century, especially in Southwest Asia and region surrounding Suez Canal; Britain promised creation of Arab state, but reneged and entered secret agreement with France to partition region—created Palestine, Transjordan, and Iraq; other regions, such as Libya, less influenced by European colonialism; Iran (once called Persia) and Turkey never directly occupied

    C. **Decolonization and Independence:** European withdrawal in North Africa before World War II, but mostly through 1950s; many countries maintained political and economic ties; Southwest Asia gained independence between 1930 and 1971, but imposed colonial-era boundaries continue to shape regional geopolitics

    D. **Modern Geopolitical Issues:** political boundaries certain to change as result of ongoing negotiated settlements or political conflict

        a. **Across North Africa:** varied settings threaten region's political stability, such as Islamist political movements; Sudan faces most daunting political issues

        b. **Arab-Israeli Conflict:** creation of Jewish state of Israel in 1948 produced enduring zone of cultural and political tensions; legacy of 1917 Balfour Declaration, which pledged to create Jewish homeland—realized in 1948 with UN division of region into Jewish and Arab Palestinian states; wars in 1948, 1956, 1967, and 1973 led to expansion of Israel; ongoing attempts to negotiate settlement; Israel construction of security barrier; political fragmentation of Palestinians contribute to uncertainty of region; instability also continues along Israel's border with Lebanon

        c. **Devastated Iraq:** multinational state created during colonial era, carved out of British empire in 1932; religious tensions between Sunnis, Shiites, and Kurds; wars in 1980s, 1990s, and 21ˢᵗ century

        d. **Instability in the Arabian Peninsula:** Saudi Arabia, nominal ally of United States, led by conservative monarchy and unwilling to promote democratic reforms; possibility of sponsoring anti-American groups and terrorist organizations

        e. **Tensions in Turkey:** political tension between pro-Western elements that seek membership in European Union and anti-Western elements; continuing tensions with Greece and Cyprus

        f. **Iranian Geopolitics:** Iran supports Shiite Islamist elements throughout region; continues to threaten Israel; uncertainty of ongoing nuclear development program

    E. **An Uncertain Political Future:** few areas pose more geopolitical questions than this region; region retains strategic global importance, especially because of growth of world's petroleum economy; also rise of Islamic fundamentalism and Islamist political movements; key question is what geopolitical role the United States will play

        a. Four interrelated issues and U.S. involvement: (1) U.S. involvement in Iraq still important; (2) relationship with oil-rich Saudi Arabia; (3) growing geopolitical presence of Iran; and (4) recent events in Israel, Gaza, and West Bank

6. **Economic and Social Development:** wide gap between rich and poor; some countries with rich reserves of petroleum and natural gas, other nations among least developed in world; persistent political instability also contributes to region's economic problems

    A. **The Geography of Fossil Fuels:** uneven distribution of these resources; world's largest concentrations within Arabian-Iranian sedimentary basin; all states bordering Persian Gulf benefit from oil and gas deposits; second important zone includes eastern Algeria, northern and central

Libya, and various parts of Egypt; third zone includes Sudan; other states, notably Israel, Jordan, and Lebanon, lie outside favored zones

B. **Regional Economic Patterns:** remarkable economic differences throughout region

    a. **Higher-Income Oil Exporters:** richest countries owe wealth to oil reserves, including Saudi Arabia, Kuwait, Qatar, Bahrain, and United Arab Emirates; able to invest in transportation networks, urban centers, schools, medical facilities, low-cost housing, modernized agriculture; problems remain, in part because of fluctuations in world oil market; some countries rapidly depleting reserves (e.g., Bahrain and Oman)

    b. **Lower-Income Oil Exporters:** secondary players in oil trade, hampered also by political and economic problems (e.g., Algeria, Iraq, and Iran)

    c. **Prospering Without Oil:** some countries developing economically without oil reserves, including Israel (large investments, productive agricultural and industrial base, and global center for high-tech computer and telecommunications); and Turkey (diversified economy, productive agriculture and industry, and tourism)

    d. **Regional Patterns of Poverty:** Poorer countries, including Sudan, Yemen, Egypt; Sudan beset with civil war; Egypt suffers also from brain drain; Yemen as poorest country, with economy based on marginally productive subsistence agriculture

C. **A Woman's Changing World:** role of women in largely Islamic region remains a major social issue; female labor participation rates among lowest in world; large gaps typically exist between male and female literacy; few women allowed to work outside home in some conservative parts of region

    a. Women's roles are changing in some places; in Algeria, 70 percent of lawyers and 60 percent of judges are women, and majority of university students are women; in Sudan and Saudi Arabia, education is segregated but available

D. **Global Economic Relationships:** region has close economic ties with the world; oil and gas as critical commodity relationships; tourism and manufacturing trade also important

    a. **OPEC's Changing Fortunes:** OPEC retains influence; many oil-producing countries, however, willing to form partnerships with foreign corporations, thus accelerating economic integration; other economic activities contribute to global integration, such as Turkey's export of textiles, food products, and manufactured goods

    b. **Regional and International Linkages:** future interconnections depend on cooperative economic initiatives; relations with European Union are crucial; most Arab countries, though, wary of too much European influence; in 2005, the Greater Arab Free Trade Area (GAFTA) was created, designed to eliminate all intraregional trade barriers and spur economic cooperation

    c. **Geography of Tourism:** tourists are another link to global economy; traditional magnets such as ancient historical and religious sites remain important; new demand for other recreational spots, including ecotourism; problems related to tourism include environmental damage and damage to some archaeological and sacred sites; also problem of local underclass developing

Summary

- The region has played a central role in world history and in processes of globalization; serves as birthplace for some of world's earliest urban civilizations; three of world's major religions emerged from region
- Many countries within region suffer from significant environmental challenges
- Region remains hearth of Christianity, Judaism, and Islam
- Political conflicts have disrupted economic development
- Abundant reserves of oil and natural gas, coupled with global economy's reliance on fossil fuels, means that region will remain important

Research or Term Paper Ideas

- Conduct library research on both Islamic fundamentalism and Christian fundamentalism. How are these two movements similar? How are they different? What factors have led to the popularity of both of these two movements?

- Conduct library research on either the Kurds or the Palestinians, two groups of people in this region who do not have a country. What characterizes these peoples as members of distinctive nations? As members of a landless nation, how have they been treated? What are the prospects that they will eventually acquire a homeland?

- Throughout the 1970s and 1980s, many oil-rich countries in the region were able to invest heavily in infrastructure projects (e.g., transportation networks, hospitals, airports, and hotels). Most of these countries, however, relied on foreign workers to both construct and later staff these. Conduct research on the global patterns of labor migration that began in the 1970s and continue to this day. From what countries were migrant workers recruited? How have these foreign workers been treated?

- Turkey has been involved in long-term discussions with the European Union to become a member of this trade bloc. Conduct research to learn why some groups within Turkey want to join the EU. What groups in Turkey oppose EU membership? What are the advantages or disadvantages to Turkey if the country joins the EU? Which side would you support?

Practice Quizzes

Answers appear at the end of this book

Vocabulary Matching: Match the term to its definition.

A. Culture hearth
B. Domestication
C. Exotic rivers
D. Fossil water
E. Hydropolitics
F. Maghreb
G. Medina
H. Monotheism

I. OPEC
J. Ottoman Empire
K. Palestinian Authority
L. Pastoral nomadism
M. Shiites (Shi'a)
N. Suez Canal
O. Sunnis
P. Theocratic state

1. _____   Water supplies stored underground during earlier and wetter climatic periods.

2. _____   A river that issues from a humid area and flows into a dry area otherwise lacking streams.

3. _____   A state in which religious leaders set and guide policy.

4. _____   An area of historical cultural innovation (for example, domestication of plants or animals).

5. _____   Muslims who practice one of the two main branches of Islam; they favored passing on power within Muhammed's own family.

6. _____   Muslims who practice one of the two main branches of Islam; they favored passing on power through established clergy.

7. _____   The interplay of water resource issues and politics.

8. _____   A traditional form of subsistence agriculture in which practitioners depend on the seasonal movement of livestock for a large part of their livelihood.

9. _____   A man-made waterway that linked the Mediterranean to the Red Sea in 1869; it was engineered by the British.

10. _____   In North Africa, the region (meaning "western island") that includes the countries of Morocco, Algeria, Tunisia.

11. _____   This occurs when plants and animals are purposefully selected and bred for their desirable characteristics.

12. _____   Belief in one God.

13. _____   A quasi-government body that represents Palestinian interests in the West Bank and Gaza in Israel.

14. _____   A large, Turkish-based empire that dominated large portions of southeastern Europe, North Africa, and Southwest Asia between the 16th and 19th centuries.

15. _____   An international organization of 12 oil-producing countries (formed in 1960) that attempts to influence global prices and supplies of oil; Algeria, Gabon, Indonesia, Iran, Iraq, Kuwait, Libya, Nigeria, Qatar, Saudi Arabia, UAE, and Venezuela are members.

Multiple Choice: *Choose the word or phrase that best answers the question*

1.   Why is the region of Southwest Asia and North Africa considered to be a world culture hearth?
    a.   It was a seedbed of agricultural domestication
    b.   Judaism, Christianity, and Islam were founded in the region
    c.   Language originated in this region
    d.   A and B above
    e.   A and C above

2.   What is the major reason for environmental problems in Southwest Asia and North Africa?
    a.   Domination of heavy industry in the region
    b.   High population density
    c.   The petroleum industry
    d.   The long history of human settlement in the region
    e.   All of the above

3.   What causes salinization (the buildup of toxic salts in the soil)?
    a.   Industrial pollution
    b.   Petroleum refining
    c.   Petroleum extraction
    d.   Salt mining
    e.   Extensive use of irrigation in desert lands

4.   Which of the following statements about climate in Southwest Asia and North Africa are true?
    a.   Aridity dominates large portions of the region
    b.   Plants and animals have adapted to extreme conditions in the region
    c.   Some of the driest parts of the region are found in North Africa
    d.   A and B above
    e.   A, B, and C above

5.   Which of these is the most accurate characterization of the physiological density of Southwest Asia and North Africa?
    a.   The physiological density is high, which means that the number of people per square mile is high
    b.   The physiological density is high, which means that the number of people per unit of arable land is among the highest on Earth
    c.   The physiological density is low, meaning that the number of livestock is relatively low compared to the number of people
    d.   The physiological density is low, meaning that there is ample farmland to support the population
    e.   The physiological density is moderate, meaning that there is a modest number of people per square kilometer

6.   Which of the following are among the most common livelihoods in Southwest Asia and North Africa?
       a.   Irrigated agriculture along exotic rivers
       b.   Oasis agriculture
       c.   Pastoral nomadism
       d.   A and B above
       e.   A, B, and C above

7.   Which of the following statements about cities in Southwest Asia and North Africa are true?
       a.   European colonialism added another layer to the urban landscape features
       b.   Many cities reflect Islamic influences
       c.   Modern popular culture has significantly altered traditional architecture throughout the region
       d.   A and B above
       e.   A and C above

8.   What pattern of migration is found in Southwest Asia and North Africa?
       a.   Large numbers of workers are migrating within the region to areas with job opportunities
       b.   Political forces and war have encouraged migration
       c.   Residents of the region are migrating to other parts of the world, especially to Europe
       d.   A and B
       e.   A, B, and C above

9.   What kind of government does Iran have?
       a.   Theocracy
       b.   Representative democracy
       c.   Democratic Republic
       d.   Constitutional monarchy
       e.   Communist dictatorship

10.   Followers of which religion form the majority population in nearly all of the countries of Southwest Asia and North Africa?
       a.   Buddhism
       b.   Coptic Christianity
       c.   Islam
       d.   Judaism
       e.   Orthodox Christianity

11.   What was the major reason for the late arrival of European colonization of Southwest Asia and North Africa?
       a.   The domination of the region by the Ottoman Empire (Turks)
       b.   Military resistance to European countries by Persia
       c.   The difficult physical geography of the region
       d.   The lack of valuable resources in the region
       e.   The presence of serious illnesses in the region and Europeans' lack of resistance (as in Africa)

12.   What country of Southwest Asia and North Africa is the only one that does not have a majority
        Muslim population?
        a.   Egypt
        b.   Israel
        c.   Libya
        d.   Morocco
        e.   Turkey

13.   Which country of Southwest Asia and North Africa functioned as a democracy for many years?
        a.   Israel
        b.   Saudi Arabia
        c.   Turkey
        d.   A and B above
        e.   A and C above

14.   To what do the most prosperous countries of Southwest Asia and North Africa owe their wealth?
        a.   Massive oil reserves
        b.   Information technology
        c.   Industry
        d.   Banking services
        e.   Export manufacturing

15.   What is the only country in Southwest Asia and North Africa to seek membership in the European
        Union?
        a.   Iraq
        b.   Israel
        c.   Libya
        d.   Turkey
        e.   United Arab Emirates

Summary of Southwest Asia and North Africa

Total Population of SW Asia and North Africa: About 542 million

Population Indicators for SW and North Africa

	Highest (country and value)	Region Average	Lowest (country and value)
Population 2010 (millions)	Egypt: 80.4	22.8	Western Sahara: 0.5
Density per sq km	Bahrain: 1,807	198	Western Sahara: 2
RNI	Yemen: 3.0	1.9	Qatar: 0.8
TFR	Yemen: 5.5	3.0	Iran; Qatar: 1.8
Percent Urban	Bahrain; Qatar: 100%	73%	Yemen: 29%
Percent < 15	Yemen: 45%	31%	Qatar: 15%
Percent > 65	Israel; Lebanon: 10%	4%	Qatar; UAE: 1%
Net Migration (per 1000;2000-05)	Qatar: 93.9	6.9	Iraq: -3.9

Development Indicators for SW Asia and North Africa

	Highest (country and value)	Region Average	Lowest (country and value)
GNI per capita/PPP 2008	Kuwait: $53,430	$14,852	Yemen: $2,220
GDP Avg. Annual Growth (2000-2008)	Qatar: 9.0%	4.38%	Iraq: -11.4%
Human Development Index (2007)	Israel: 0.935	0.786	Yemen: .575 Sudan: 0.531
Percent Living below $2/day	Yemen: 46.6%	16.9%	Jordan: 3.5%
Life Expectancy 2010	Israel: 81	74	Sudan: 58 Western Sahara: 60
< 5 Mortality 2008	Sudan: 109 per 1000 Yemen: 69 per 1000	28 per 1000	Israel: 5 per 1000
Gender Equity 2008	Qatar: 120	101	Morrocco: 88

Chapter 8

Europe

Learning Objectives

➢ This chapter introduces Europe, one of the world's most densely settled modern regions.

➢ An important objective is to understand the dynamics of nationalism and the numerous conflicts that besieged this region throughout the twentieth century.

➢ The chapter sheds light on the changing dynamics that have resulted in the recent formation of the European Union and the evolution of a common currency.

➢ The following concepts and models are introduced and elaborated: Balkinization, command economy, European Union, privatization, and secularization.

Chapter Outline

1. **Introduction:** more than half a billion people reside in 41 countries of Europe; the region exhibits remarkable cultural diversity; despite centuries of conflict, there now exists a spirit of cooperation, marked by the supranational organization known as the European Union

2. **Environmental Geography**
   A. **Environmental Diversity:** four factors contribute
      a. Complex geology
      b. Europe's latitudinal extent
      c. Moderating influences of water bodies
      d. Long history of human settlement
   B. **Landforms and Landscape Regions:** four general topographic regions are identified
      a. **European Lowland:** also known as North European Plain; high population density, intensive agriculture, large cities, major industrial regions; many of Europe's major rivers flow across lowland (Rhine, Loire, Thames, and Elbe); several important ports (London, Le Havre, Rotterdam, and Hamburg)
      b. **Alpine Mountain System:** topographic spine of Europe; stretches east to west from Atlantic to Black Sea and runs southeast to the Mediterranean; mountain ranges with local names: Pyrenees (between Spain and France), Alps (from France to eastern Austria), Apennines (spine of Italy), Carpathians (from eastern Austria to borders of Romania and Serbia), Dinaric Alps, Balkan Ranges
      c. **Central Uplands:** between Alps and European Lowland in France and Germany; contain many raw materials for Europe's industrial regions
      d. **Western Highlands:** define the western edge of European subcontinent, from Portugal in south through British Isles, to highlands of Norway, Sweden, and Finland; specific place-names; characterized by U-shaped glaciated valleys and fjords (flooded valleys); Fenno-Scandian Shield of Sweden and northern Finland—some of the oldest rocks in world
   C. **Europe's Climates:** three principal climates
      a. **Maritime Climate:** along Atlantic coast; moderating influence of sea
      b. **Continental Climate:** interior of Europe; more extreme seasonal differences; transition from marine to continental located along Rhine River border of France and Germany
      c. **Mediterranean Climate:** distinct dry season in summer; located along Mediterranean; seasonal drought can be problem
   D. **Seas, Rivers, and Ports:** Europe remains a maritime region; even landlocked countries (e.g., Austria, Hungary, Serbia, and Czech Republic) have access to sea via interconnected rivers and canals

    a. **Europe's Ring of Seas:** major seas encircle Europe—North Sea, Mediterranean Sea, Aegean Sea, Black Sea, Adriatic Sea, and Baltic Sea; also Atlantic Ocean; key strategic channels and straits, such as the English Channel, Bosporus Strait, and Strait of Gibraltar

    b. **Rivers and Ports:** Europe is a region of navigable rivers connected by a system of canals and locks; Danube—Europe's longest river; major ports located at mouths of most western European rivers—transshipment points for inland waterways, rail and truck transportation

  E. **Environmental Issues:** long history of environmental problems, but region is increasingly effective in addressing problems with regional solution; European Union is world leader in recycling, waste management, reduced energy use, and sustainable resource use; conditions are worse in Eastern Europe—partly as a result of Soviet economic planning—but future appears more positive

    a. **Global Warming in Europe:** indications include dwindling sea ice, melting glaciers, sparse snow cover in arctic Scandinavia, increasing droughts in Mediterranean, and rising sea levels; Europe has become world leader in addressing global warming

      i. **Europe and Kyoto Procol:** EU's philosophy: regional action to solving environmental problems superior to that of individual state action; EU proposed collective goal of greenhouse gas emission—to permit growth of poorer countries with more developed states making emission reductions; umbrella remains in place

      ii. **EU's Emission Trading Scheme:** in 2005, EU initiated carbon trading scheme; cap and trade system to make business more expensive for those who pollute while rewarding those industries that stay under carbon quota

      iii. **Results to Date:** not on target to meet 2012 goal of 8 percent reduction; failure to anticipate growth in truck transport, coupled with excessive emission from industry in Spain and new EU countries of eastern Europe; western Europe has reduced emissions and increased energy generation from wind and solar power

3. **Population and Settlement:** population densities higher in historical industrial core of western Europe; lower in eastern and northern periphery; urban-industrial core has low natural growth rates, and target for immigration

  A. **Natural Growth:** death rate in many countries exceeds birth rate—negative growth, suggesting fifth state of demographic transition model—post-industrial phase in which population growth falls below replacement levels

    a. **Low-Birth-Rate Countries**—attempting to increase rates of natural growth; most EU countries have liberal maternity benefits; some give cash awards to couples when they marry and have children

  B. **Migration To and Within Europe:** migration is one of most important population issues; widespread resistance to unlimited migration into Europe; controversial topic

    a. During 1960s, many countries encouraged immigration for labor needs; now Europe suffers from economic recession and concerned over continued immigration—especially from former colonies; political and economic troubles in Eastern Europe and Russian domain generated more migration to Europe

    b. **Fortress Europe:** opponents of large-scale immigration encourage more regulation, establishing "defensive perimeter," with "hard" borders along peripheral borders; maintain "soft" borders within Europe for ease of European movement; Schengen Agreement (1985): goal of gradual reduction of border formalities for travelers moving between EU countries

  C. **Landscapes of Urban Europe:** high level of urbanization; gradient from highly urbanized European heartland to less-urbanized periphery reinforces differences between affluent industrial core and more rural, less developed periphery

  D. **The Past in the Present:** cities as mosaic of historical and modern landscapes; three historical eras dominate: *medieval landscape* of narrow, winding streets, crowded three- or four-story masonry buildings with little set-back from streets; *Renaissance-Baroque* of more open and spacious areas, ceremonial buildings; *industrial*, with clustered factories, rail terminals and train stations; also post-World War II planning, with suburban sprawl, green spaces, and well-developed public transportation systems

4. **Cultural Coherence and Diversity:** mosaic of languages, customs, religions; strong regional identities; European cultures play leading role in globalization
    A. **Geographies of Language:** millions of Europeans learn multiple languages
        a. **Germanic Languages:** dominate Europe north of the Alps; German is dominant in Germany, Austria, Liechtenstein, Luxembourg, and eastern Switzerland; English is second largest Germanic language
        b. **Romance Languages:** evolved from everyday Latin (e.g., French, Spanish, and Italian); many, such as French and Spanish, have strong regional dialects
        c. **Slavic Language Family:** largest European sub-family of Indo-European languages; traditional split between northern groups (Polish, Czech, and Slovakian) and southern (Hungarian and Romanian); use of two alphabets—those with strong Roman Catholic heritage (e.g., Poland and Czech Republic) use Latin alphabet, while those with ties to Orthodox church (e.g., Bulgaria, Macedonia, Serbia) use Greek-derived Cyrillic alphabet
    B. **Geographies of Religion, Past and Present:** many of today's ethnic tensions result from historical religious events
        a. **Schism between Western and Eastern Christianity:** eastern church split in 1054, with further splintering of Greek Orthodox, Bulgarian Orthodox, Russian Orthodox; eastern Christianity followed Cyrillic alphabet
        b. **Protestant Revolt:** another split within Christianity occurred with split between Catholics and Protestants during 16th century
        c. **Conflicts with Islam:** historical conflicts between two great religions (e.g., crusades); tensions today related to immigrant Muslim populations
        d. **Geography of Judaism:** Europe as long-time difficult home for Jews; eastern Europe throughout Middle Ages known as *Jewish Pale*—large-scale settlement of Jews in what is today eastern Poland, Belarus, western Ukraine, and northern Romania; prior to large-scale emigration (c. 1890), Europe, and especially Jewish Pale, was home to 90 percent of world's Jewish population; Holocaust led to the death of 6 million Jews
        e. **Patterns of Contemporary Religion:** Catholics mostly in southern half of region; Protestants in north; there has been noticeable decline in church attendance—secularization, referring to movement away from importance of organized religion; numerous Muslims, mostly immigrants from Africa and southwestern Asia
    C. **European Culture in Global Context:** Europe undergoing profound cultural changes; complicated interaction between globalization and Europe's internal agenda of political and economic integration
        a. **Migrants and Culture:** historically, Europe spread its culture via colonialism; now Europe is experiencing reverse flow with migrants bringing their cultures to region; establishment of ethnic clusters; political landscape reveals emergence of far-right, nationalistic parties as backlash to immigration
5. **Geopolitical Framework:** dense fabric of 41 independent states in relatively small area; no other world region demonstrates such a mosaic of geopolitical divisions; two world wars forced redrawing of political boundaries in 20th century; ongoing divisions and fragmentations
    A. **Redrawing the Map of Europe Through War:** two world wars radically reshaped geopolitical maps; new states were born; old kingdoms and empires overthrown; emergence of *irredentism*--state policies for reclaiming lost territory and peoples; during 1930s and 1940s, three competing political ideologies underpinned changes: Western democracy and capitalism, communism, and fascist totalitarianism
    B. **Divided Europe, East and West:** from 1945 until 1990, Europe divided into two geopolitical and economic blocs: democratic and capitalistic West and communistic East; separated by *Iron Curtain*

    a. **Cold War Geography:** seeds of Cold War planted in 1945 (Yalta Conference) with decision for main powers (United Kingdom, United States, and Soviet Union) to shape/divide postwar Europe; Soviet's desired an establishment of a *buffer zone*, a series of territories dominated politically and economically by Soviet Union, to serve as a barrier from the West

    b. **Cold War Thaw:** symbolic end of Cold War on November 9, 1989, with demolition of the Berlin Wall; by October 1990, East and West Germany reunited; other former Soviet satellites achieve independence; most were peaceful revolutions with free elections (exception being Romania)

  C. **The Balkans:** troublesome area, spawns term "balkanization" to describe geopolitical processes of small-scale independence movements based upon ethnic fault lines; in 1990s, fragmentation of former Yugoslavia

    a. **Geo-Historical Background:** following World War I, collapse of Austro-Hungarian and Ottoman Turk Empires; development of multi-ethnic state of Yugoslavia—a political entity composed of three major religions, handful of languages, at least eight different ethnic groups, and two distinct alphabets; geopolitical stability unraveled in 1990s

    b. **Balkan Wars of Independence:** Slovenia and Croatia declared independence in 1991, Macedonia in 1992, Bosnia and Herzegovina in 1992; Serbians attempt to maintain unified territory; Dayton Peace Accords in 1995; Kosovo another key flashpoint—finally ruled independent by UN in 2010, but Serbia disputes ruling

    c. **Moving Toward Stability:** Serbian diplomatic relations with United States and other countries have been normalized

6. **Economic and Social Development:** Europe "invented" modern economic system of industrial capitalism; movement toward economic integration over past 50 years—providing new model for regional cooperation; throughout western Europe, unprecedented levels of social development

  A. **Europe's Industrial Revolution**: two fundamental changes spurred industrialization—machines replaced human labor, and inanimate energy sources power new machines; occurred over a century, led by England (between 1730 and 1850); spread throughout rest of world

    a. **Centers of Change:** England's textile industry, located on flanks of Pennine Mountains, was center of industrial innovation

    b. **Development of Industrial Regions in Continental Europe**: by the 1820s, the first industrial districts appeared, mostly near coal fields, such as Sambre-Meuse region straddling French–Belgian border; by the late nineteenth century, the dominant industrial area in Europe was Ruhr district in northwestern Germany

  B. **Rebuilding Postwar Europe:**

    a. **ECSC and EEC:** in 1950s, attempt to promote economic integration led to establishment first of European Coal and Steel Community, followed by establishment of European Economic Community

    b. **From the EEC to the EU:** in 1965 Brussels Treaty, added a political dimension to integration—the EEC becomes the European Community; in 1991 following Treaty of Maastricht, the EC became the European Union; EU concerned now also with supranational affairs, such as common foreign policies and mutual security agreements; EU continues to add members, including former Soviet-controlled communist countries; others continue to apply for membership

    c. **Euroland:** from 1999, much of Europe moved to a common currency, the *Euro*; initially 15 EU member states formed the European Monetary Union (EMU); Euroland members increased efficiency and competitiveness of domestic and international business through common currency; some EU members (e.g., United Kingdom) have reservations about joining EMU; membership remains controversial

  C. **Economic Integration, Disintegration, and Transition:** historically, eastern Europe less developed economically; subregion not as rich in natural resources; few resources usually exploited by outside interests (e.g., Ottoman Empire, Germany, and Soviet Union); Soviet-

dominated economic planning (command economy) dominated region from 1945 to 1990; centralized system collapsed in 1991, leading to economic, political, and social chaos
   a. **Results of Soviet Economic Planning:** results were mixed; most farmers resisted national ownership of agriculture and most productive land remained in private hands; but food shortages throughout were common; industrialization was developed, but overall region experiences shortages of consumer goods
   b. **Transition and Change since 1991:** collapse of Soviet Union led to forced economic integration of region; troubled period of economic transition; compounded problems with Russia's curtailment of cheap natural gas and petroleum; many eastern European countries redirect economies away from Russia and toward EU
D. **Regional Disparities in eastern Europe:** most eastern European countries have joined, or are in process of joining, EU; some are doing better than others (Slovenia and Czech Republic doing well, less so for Albania)

Summary

- Europe has made great progress toward environmental protection
- Europe continues to face challenges related to population and migration
- Political and cultural tensions remain high regarding migration
- Major changes have taken place in Europe's geopolitical structure throughout twentieth century
- European integration is facing increasing challenges resultant from economic recession

Research or Term Paper Ideas

- Compare the emergence and progression of the European Union with the establishment of regional organization in sub-Saharan Africa. How do they differ? What are the various goals and objectives? Why has the European Union fared better than those in sub-Saharan Africa?

- What are the social, economic, and political issues related to immigration into Europe?

- Two regions of Europe are already threatened by rising sea levels: the Dutch coastline and the Italian coastline around Venice. Learn more about these two locations and what the respective governments have done to mitigate the problem

Practice Quizzes

Answers appear at the end of this book

Vocabulary Matching: Match the term to its definition.

A. Balkanization
B. Berlin Wall
C. Cold War
D. Euroland
E. European Union
F. Fjord
G. Guest workers
H. Iron Curtain

I. Irredentism
J. Marine west coast climate
K. Medieval landscape
L. Mediterranean climate
M. Moraine
N. Renaissance-Baroque landscape
O. Schengen Agreement
P. Secularization

1. _____ The widespread movement away from the historically prominent religions of Europe.

2. _____ Individuals from Europe's agricultural periphery invited to labor in Germany, France, Sweden, and Switzerland during chronic labor shortages in Europe's boom years (1950s–1970s).

3. _____ Piles of glacial debris that were left on plains areas in Germany and Poland.

4. _____ Climate type characterized by a distinct dry season during the summer, which results from the warm-season expansion of the Azores high-pressure area.

5. _____ A European landscape with narrow, winding streets, crowded with 3–4 story masonry buildings with little setback from the street.

6. _____ Climate type with no winter months averaging below freezing, although cold, rain, sleet, and the occasional blizzard are common.

7. _____ The ideological struggle between the United States and the Soviet Union that lasted from 1946 till 1991.

8. _____ U-shaped glaciated valleys that have become flooded valley inlets along the coastline of Norway.

9. _____ A European landscape built from 1500–1800 with a spacious landscape, expansive ceremonial buildings, public open and green spaces, and wide boulevards lined with elaborate residences.

10. _____ A supranational organization made up of 25 countries, joined together in an agenda of economic, political, and cultural integration.

11. _____ States within the European Union that gave up their national currencies and adopted the official currency of the EU, the Euro.

12. _____   The fragmented geopolitical processes involved with small-scale independence movements and the phenomenon of mini-nationalism as it develops along ethnic fault lines.

13. _____   The 1985 agreement between some (but not all) European Union member countries to reduce border formalities in order to facilitate free movement of citizens between member countries of this group.

14. _____   State policies for reclaiming lost territory (real or imagined) beyond borders inhabited by people of the same ethnicity.

15. _____   A term coined by British leader Winston Churchill during the Cold War that defined the western border of Soviet power in Europe.

Multiple Choice: *Choose the word or phrase that best answers the question.*

1.   Which of the following climates is absent from Europe?
   a.   Continental midlatitude
   b.   Tropical
   c.   Mild midlatitude
   d.   Mediterranean
   e.   Highland

2.   Which of the following statements about bodies of water in and around Europe are true?
   a.   The seas surrounding Europe are connected to each other through narrow straits
   b.   Five major seas circle Europe
   c.   Major ports are found at the mouth of most rivers of Western Europe
   d.   A and B above
   e.   A, B, and C above

3.   Which of the following is responsible for the environmental problems facing Europe today?
   a.   Industry and urbanization
   b.   Resource extraction
   c.   Long history of agriculture
   d.   A and B above
   e.   A, B, and C above

4.   At which stage of the Demographic Transition is Europe?
   a.   First (high birth rate, high death rate)
   b.   Second (high birth rate, declining death rate)
   c.   Third (declining birth rate, declining death rate)
   d.   Fourth (low birth rate, low death rate)
   e.   Fifth stage (postindustrial, where population falls below replacement)

5.   A recent migrant to England is most likely to come from which of the following countries?
   a.   Algeria
   b.   Suriname
   c.   India
   d.   Vietnam
   e.   Turkey

6. Where are the "hard" borders of Europe?
   a. On the perimeter of the European Union
   b. At the edge of continental Europe
   c. Between the countries of Europe
   d. Between Western and Eastern Europe
   e. At the eastern edge of Europe, where the region meets Asia

7. Which of these statements about language in Europe are correct?
   a. As their first tongue, 90 percent of Europeans speak a Germanic, a Romance, or a Slavic language, but many speak two or more languages from the region
   b. English is the official language of Europe
   c. Some small ethnic groups work hard to preserve their cultural identities and languages
   d. A and B are correct
   e. A and C are correct

8. What variation of Christianity is found in Europe?
   a. Roman Catholicism
   b. Orthodox Christianity
   c. Protestantism
   d. A and B above
   e. A, B, and C above

9. What is Europe's major geopolitical issue today, at the beginning of the 21st century?
   a. Integration of eastern and western Europe into the European Union
   b. The wars in Iraq and Afghanistan
   c. Region-wide environmental legislation
   d. Resolution of border disputes between European countries
   e. All of the above

10. Which of the following European regions was part of the buffer zone for the former Soviet Union during the Cold War?
    a. Iberian Peninsula
    b. Baltic states
    c. Scandinavia
    d. European lowland
    e. Alpine region

11. Why did the Soviet Union build the Berlin Wall?
    a. To defend against attacks on its western boundary
    b. To serve as a routine border checkpoint
    c. To stop Germans from fleeing from East Germany to West Germany
    d. A and B above
    e. A, B, and C above

12. In which part of Europe has ethnic tension been so high for so long that its name has become a part of the word to describe the fragmented geopolitical process involved with small-scale independence movements and mini-nationalism as it develops along ethnic fault lines?
    a. Iberia
    b. Scandinavia
    c. The Balkans
    d. The Baltic States
    e. Benelux

13. When the first industrial districts appeared in continental Europe in the 1820s, which of these was the key determinant of location?
    a. Coalfields
    b. Forests
    c. Oil wells
    d. Ports
    e. Rivers and water wheels

14. Eastern Europe's sluggish economy and relatively poor economic performance are the result of its past history with which foreign country?
    a. China
    b. Japan
    c. Russia
    d. United States
    e. West Germany

15. What important milestone did the European Union celebrate in 2007?
    a. The introduction of the Euro
    b. Admission of Turkey into the EU
    c. Its 50th anniversary
    d. The ratification of the Schengen Agreement
    e. A and D above

Summary of Europe

Total Population of Europe: About 531 million

Population Indicators for Europe

	Highest (country and value)	Region Average	Lowest (country and value)
Population 2010 (millions)	Germany: 81.6	13.73	San Marino: 0.03
Density per sq km	Monaco: 34,835	1,075	Iceland: 3
RNI	Iceland: 0.9	1.51	Servia: -0.5
TFR	Iceland; Ireland: 2.1	1.52	Bosnia and Herzegovina; San Marino: 1.2
Percent Urban	Monaco: 100%	69%	Liechtenstein: 15%
Percent < 15	Monaco: 13%	30.76%	Albania: 25%
Percent > 65	Monaco: 24%	16%	Albania: 9%
Net Migration (per 1000;2000-05)	Iceland: 12.8	2.2	Lithuania: -6.0

Development Indicators for Europe

	Highest (country and value)	Region Average	Lowest (country and value)
GNI per capita/PPP 2008	Norway: $59,250	$26,383	Albania: $7,520
GDP Avg. Annual Growth (2000-2008)	Latvia: 8.2%	3.9%	Portugal: 0.9%
Human Development Index (2007)	Iceland: 0.965	0.911	Bosnia and Herzegovina: 0.812
Percent Living below $2/day	Albania: 7.8%	0.4%	Most countries: <2%
Life Expectancy 2010	Switzerland; Italy; Montenegro: 82	78	Lithuania: 72
< 5 Mortality 2008	Finland; Sweden: 3 per 1000	6 per 1000	Bosnia and Herzogovina: 15 per 1000
Gender Equity 2008	Spain; Ireland: 103	100	Austria; Switzerland; Greece: 97

Chapter 9

The Russian Domain

Learning Objectives

> ➢ This chapter introduces the region formerly dominated by the Union of Soviet Socialist Republics (U.S.S.R.).
> ➢ The interrelations between the former Soviet Union and the United States are stressed, highlighting how Cold War political tensions influenced the political geography of the region.
> ➢ The unique aspects of the Russian domain's northern latitudinal location are stressed.
> ➢ You will gain insight into the differences between 'command economies' and more democratic, capitalist economies.
> ➢ The following concepts and models will be developed: Cold War, permafrost, Russification, Glasnost and perestroika, command economy.

Chapter Outline

1. **Introduction:** The Russian domain extends across the northern half of Eurasia; it includes Russia, Ukraine, Moldova, Georgia, and Armenia
   A. The regional definition reflects the changing political and cultural map since the breakup of the Soviet Union; the term *domain* suggests a persisting Russian influence within the five other states
   B. Two significant areas that once were included in the Soviet Union are now considered elsewhere: the mostly Muslim Republics of Central Asia and the Caucasus, and the Baltic Republics (now considered part of Europe)
   C. Globalization is shaping the Russian domain in complex ways; the now-independent republics are carving out new regional and global relationships; social, political, and economic tensions remain

2. **Environmental Geography:** region exhibits some of the world's most severe environmental degradation; approximately two-thirds of the Russian population live in an environment harmful to their health; legacy of seven decades of Soviet industrialization—careless mining, oil drilling, nuclear testing; environmental degradation of Siberian forests
   A. **Air and Water Pollution:** poor air quality plagues much of the region, resultant from clusters of industrial processing and manufacturing, often with minimal environmental controls; led to significant contamination of the region's forests; degraded water also a significant problem; many seas and rivers heavily polluted
   B. **Nuclear Threat:** nuclear weapons and nuclear energy programs expanded rapidly since 1950s; environmental safety was often ignored; many areas suffer from nuclear fallout following above-ground testing; Russian Arctic poisoned from unregulated dumping of nuclear wastes; nuclear pollution especially pronounced in northern Ukraine
   C. **Post-Soviet Challenges:** end of Soviet rule has had mixed consequences for environment; on the positive side, advanced pollution control equipment being used; nuclear warhead storage facilities consolidated; and environmental consciousness is growing; on negative side, Russia depends heavily on natural resources to promote economic growth; state-run petroleum and natural-gas companies are politically powerful; Siberian forests rapidly diminishing; also crisis of wildlife extinctions

D. **A Diverse Physical Setting:** northern latitudinal position is critical to understand geographies of climate, vegetation, and agriculture; exhibits seasonal temperature extremes; short growing season limits opportunities for agriculture; except for subtropical zone near Black Sea, region dominated by continental climate

    a. **European West:** consists of European Russia, Belarus, Ukraine; various river systems linked by canals—facilitates interconnectivity; key rivers include Dnieper, Don, West and North Dvina, and Volga

        i. Three distinct environments for agriculture: north of Moscow and St. Petersburg—poor soils and cold temperatures limit farming; Belarus and central Russia—longer growing seasons but acidic soils limit farming, though more diversified; southern portion more conducive to agriculture—longer growing season, better soils, important for commercial wheat, corn, and sugar beets

    b. **Ural Mountains and Siberia:** mark European Russia's eastern edge; mountains not exceptionally high, but contain valuable mineral resources; Siberian landscape, with key rivers of Ob, Yenisey, Lena—consists mostly of tundra vegetation, associated with *permafrost* (cold-climate condition of unstable, seasonally frozen ground)—this limits growth of vegetation and causes problems for railroad construction; interior region also of Russian *taiga* (coniferous forest zone)

    c. **Russian Far East:** distinctive sub-region proximate to Pacific Ocean; fertile river valleys; ecological transition zone from continental climates of Siberian interior and seasonal monsoon of East Asia

    d. **The Caucasus and Transcaucasia:** region's extreme south, dominated by Caucasus Mountains between Black and Caspian Seas; fragile environment; farther south in Transcaucasia are Georgia and Armenia; the former is important producer of subtropical fruits, vegetables, flowers, and wine

E. **Global Warming in the Russian Domain:** a world region that might benefit from warmer global climate; however, human–environment relationship is complex

    a. **Potential Benefits:** northern limit of cereal cultivation in northwestern Russia would be extended; areas once in tundra vegetation able to support boreal forests; less severe winters provide more opportunities for energy and mineral development; less sea ice in Arctic Ocean and Barents Sea means easier navigation, more high-latitude commerce, easier drilling for oil and gas; Siberian rivers (Ob and Yenisey) may be better suited for commerce

    b. **Potential Hazards:** hotter summers might increase the risk of wildfires; ecologically sensitive arctic and sub-arctic ecosystems being disrupted—possible extinctions of some wildlife; thawing of the Siberian permafrost will lead to huge release of carbon, thus augmenting global warming

3. **Population and Settlement:** Six states of Russian domain home to about 200 million people; most live in cities; population concentrated in traditional centers of European west; low birthrates and higher death rates remain critical concerns

A. **Population Distribution:** population remains heavily concentrated in west

    a. European Core: region's largest cities, biggest industrial complexes, and most productive farms located in European Core (which includes Belarus, much of Ukraine, and Russia west of Urals); especially along shores of the Baltic; other clusters oriented along lower and middle stretches of Volga River

    b. **Siberian Hinterlands:** divided into two zones of settlement, both linked to railroad lines: (1) isolated but sizable urban centers along Trans-Siberian Railroad (Omsk, Irkutsk, and Vladivostok); and (2) smaller number of settlements along Baikal-Amur Mainline Railroad

B. **Regional Migration Patterns**

    a. **Eastward Movement:** historical movement of European Russians east across Siberian frontier; pace accelerated with construction of Trans-Siberian Railroad; attractive agricultural opportunities opened; some political freedoms sought; eastward movement continued under Soviet rule

b. **Political Motives:** infilling of southern Siberian hinterland had political and economic rationale; uncounted millions also forcibly sent to *Gulag Archipelago*—vast collection of political prisons located throughout Siberia; process of *Russification*—the Soviet policy of resettling Russians into non-Russian portions of the Soviet Union, as Russians given economic and political incentives to resettle; process of Russification was geographically selective

c. **New International Movements:** in post-Soviet era, reversal of Russification; newly independent non-Russian countries imposed language and citizenship requirements—many Russians returned to Russia; since 2005, Russian government promoted repatriation program for ethnic Russians worldwide; Russia also experiencing growing immigrant population (many illegal)—increasingly from Central Asia

d. **The Urban Attraction:** Russian, Ukrainian, and Belorussian rates of urbanization comparable to industrialized capitalist countries; previously, Soviet cities grew according to government plans—different cities designed for specific industries or administrative functions; now cities prosper or not based on economic factors; those receiving foreign investment—particularly in western and southern Russia—are growing; those in Russian northeast are declining

C. **Inside the Russian City:** large Russian cities possess core area, usually with superior transportation connections; largest of cities feature extensive public spaces, monumental architecture; distinctive pattern of circular land-use zones

    a. **Mikrorayons**—large, Soviet-era housing projects, typically composed of massed blocks of standardized apartment buildings

    b. **Dacha**—elite cottage communities, found especially on Moscow's urban fringe

D. **The Demographic Crisis:** since 2005, population loss identified as key issue of national importance in Russia; similar conditions affecting other countries within region; Russian leaders propose long-term program to raise birthrates; other problems of providing health care for population

4. **Cultural Coherence and Diversity:** Slavic cultural imprint throughout region

A. **Heritage of the Russian Empire:** eastward and southward expansion of Russia

    a. **Origins of the Russian State:** Slavs originated in or near Pripyat marshes of modern Belarus; around 2,000 years ago, migrated to the east, reaching site of modern-day Moscow by AD 200 CE; Slavs intermarried with southward-moving people from Sweden (the *Varangians* or *Rus*); by AD 1000, state of Rus came into existence, with capital at Kiev (in modern-day Ukraine); Kiev-Rus state interacted with Byzantine Empire—introduction of Christianity and Cyrillic alphabet; early Russian state later split into several principalities and ruled by foreigners—Mongols and Tatars (group of Turkish-speaking peoples)

    b. **Growth of Russian Empire:** by the fourteenth century, northern Slavic people overthrow Tatar rule; expanded state; core of new Russian empire centered near eastern fringe of old state of Rus; languages in region diverged; distinctive nationalities emerge (e.g., Belorussians); Russian empire expanded in the sixteenth and eighteenth centuries; Russians allied with seminomadic Cossacks—Slavic-speaking Christians; by the seventeenth century Russian expansion went to Pacific Coast; westward expansion slower; during nineteenth century, southern expansion into Central Asia

    c. **Legacy of Empire:** Russia's ambivalent relationship with western Europe shares historical legacy of Greek culture and Christianity, but Russia long suspicious of European culture and social institutions

B. **Geographies of Language:** Slavic languages dominate the region; Russian, Belorussian, and Ukrainian are closely related

    a. Patterns in Belarus, Ukraine, and Moldova: population of Belarus mostly Belorussian—and most Belorussians live in Belarus; only about 67 percent of population in Ukraine are Ukrainian—Russian speakers make up about 25 percent; in Moldova, Romanian speakers are dominant

b. **Patterns within Russia:** Approximately 80 percent of Russia's population claim Russian linguistic identity; Finno-Ugric peoples, though small in number, dominate non-Russian north; distinctive ethnic groups scattered; Altaic speakers include Volga Tatars; Yakut peoples of northeastern Siberia, also Turkish speakers; bilingual education is common; plight of many native peoples in central and northern Siberia parallels situation in United States, Canada, and Australia

c. **Transcaucasian Languages:** bewildering variety of languages; at least three language families (Caucasian, Altaic, and Indo-European) spoken in relatively small area

C. **Geographies of Religion:** Most Russians, Belorussians, and Ukrainians share religious heritage of Eastern Orthodox Christianity; under Soviet Union, religion was severely discouraged and actively persecuted

a. **Contemporary Christianity:** since collapse of Soviet Union, region has experienced religious revival; Orthodox Church reclaiming important role as state church; other forms of Christianity present

b. **Non-Christian Traditions:** Islam is largest, with approximately 20 million adherents in region; most are Sunni Muslims, located in North Caucasus; Russia, Belarus, and Ukraine home to more than one million Jews; Buddhists present in Russian interior

D. **Russian Culture in Global Context**

a. **Soviet Days:** since 1920s, Soviet artistic productions centered on *socialist realism*, a style devoted to realistic depiction of workers harnessing the forces of nature or struggling against capitalism; traditional high arts, such as classical music and ballet continued

b. **Turn to the West:** from 1980s onward, younger generations adopting more Western-style music, dress, art, and so on; from 1991 onward, basic freedoms brought inrush of global cultural influences, including books and magazines; cultural influences also more global, with influence from East and South Asia, and Latin America

c. **The Music Scene:** MTV Russia established 1998; Western corporations linking with Russian operations; emergence of home-grown pop-music culture; some singers speaking out against human rights abuses, including domestic violence against women

d. **Revival of Russian Nationalism:** counter-trends evident, including promotion of glories of Soviet era

5. **Geopolitical Framework:** legacy of former Soviet Union weighs heavily on region; former Soviet republics still struggle to define new geopolitical identities; Russia's new global visibility has raised concerns within domain, in Europe, and in the United States

A. **Geopolitical Structure of Former Soviet Union:** Soviet Union emerged from Russian Empire in 1917 following Bolshevik (faction of Russian communists) Revolution; new socialist state established

B. **Soviet Republics and Autonomous Areas:** geopolitical administrative divisions; 15 "union republics" composed of non-Russian citizens were demarcated—theoretically, each republic retained political autonomy, but in practice, Soviet Union was a centralized state controlled from Moscow; autonomous areas—ethnic homelands located in the interior of the Soviet Union; autonomy within these "autonomous republics" was also fictitious

C. **Centralization and Expansion of Soviet State:** national autonomy within republics did not exist; Soviet state was increasingly centralized; dominated by strong leaders such as Joseph Stalin; the Stalin period, likewise witnessed geographic expansion of Soviet domination into Eastern Europe—this period marked the global *Cold War*—a situation of escalating military competition that lasted from 1948 to 1991

D. **End of the Soviet Union:** the existence of nationally based republics and autonomous areas maintained distinct cultural identities; under economic troubles, Soviet president Mikhail Gorbachev initiated policy of *glasnost* (greater openness) during the 1980s; several republics demanded outright independence; other factors contributed to downfall of Soviet Union, including failed war in Afghanistan, multiple protests against Soviet domination throughout Eastern Europe, worsening domestic economic conditions—with increasing food shortages and declining quality of life; in response, initiation of policy of *perestroika* (planned economic restructuring); by 1991, all 15 republics achieved independence and Soviet Union ceased to exist

E. **Current Geopolitical Setting:** Post-Soviet Russia and nearby independent republics have rearranged political relationships; initial effort to promote *Commonwealth of Independent States* (CIS), with most former republics joining; however, CIS mostly a forum for discussion, with no real economic or political power; Russia continues to maintain military presence in many former republics; denuclearization (return of nuclear weapons from republics to Russia) was initiated in 1990s

F. **Geopolitics in the South and West:** Russia shares cultural and political ties with Belarus and Ukraine; Belarus firmly in Russia's political orbit, while Ukraine less predictable; Moldova experiencing political tensions, especially in Transdniester region in eastern part of country; Transcaucasia remains unstable, especially Georgia, with dispute over Abkhazia and South Ossetia, two regions that border Russia; Armenia maintains close connections with Russia while building ties with United States and European Union; within Russia—ongoing centralization of authority

G. **Russian Challenge to Civil Liberties**: several public protests against abused human rights in Russia; economic uncertainties contribute to protests against central government; Russian state responds with crackdown on civil liberties

H. **The Shifting Global Setting:** regional political tensions continue; territorial disagreements with Japan over southernmost islands of Kuril Archipelago; Russia worries about expansion of North Atlantic Treaty Organization (NATO)—and especially addition of Baltic republics (Estonia, Latvia, and Lithuania); Russian leaders appear determined to reassert global political status and assert own interests in dealing with North Korea, Iran, and Venezuela

6. **Economic and Social Development:** Fluctuating economic trends—declines in 1990s, growth between 2002 and 2008, declines most recently; economic potential of Russia difficult to gauge; positives include abundant natural resources and well-educated, urbanized population; negatives include rising transportation costs in expansive region, and northern location limits agricultural productivity

A. **Legacy of the Soviet Economy:** beginning in 1920s and 1930s, imposition of centralized economic planning—state controlled production targets and industrial output; emphasized heavy, basic industries; shifted agriculture into large-scale collectives and state-controlled farms; basic infrastructure—roads, rail lines, canals, dams, communication networks—originated during this period to facilitate industry; problems during 1970s and 1980s, with agriculture and manufacturing inefficiency

B. **Post-Soviet Economy:** highly centralized, state-controlled economy replaced by mixed economy of state-run operations and private enterprise; problems of unstable currencies, corruption, and changing government policies hampered many countries in the region

   a. **Redefining Regional Economic Ties:** economic ties remain among six states, but have been redefined; less predictable flows of foreign trade; Russia's influence remains paramount

   b. **Privatization and State Control:** 1993 initiation of massive program to privatize Russian economy; opened economy to private initiative and investment; but came with few legal and financial safeguards, inviting abuses, mismanagement, and corruption

      i. Agricultural sector continues to struggle; farmers not skilled in dealing with uncertainties of market-driven agricultural economy

      ii. Privatization of service sector continuing, but long-established "informa economy" continues to flourish

      iii. Natural resources and heavy industry initially privatized, but there is increasing state control of nation's energy assets and infrastructure

   c. **Challenge of Corruption:** corruption remains widespread; cumbersome national and regional tax policies prevent efficient revenue collections; organized crime remains pervasive

C. **Ongoing Social Problems:** street crime remains high; higher rates of unemployment exist; declining social expenditures; violence against women widespread; prostitution rings prevalent in many countries, including Armenia and Moldova; health care is a major problem; alcohol-related diseases; cardiovascular disease; tuberculosis; HIV/AIDs; infant mortality; toxic environmental conditions

   D. **Growing Economic Globalization:** region remains one of the least globalized parts of more developed world

      a. **A More Globalized Consumer:** numerous consumer imports now available for people of Russian domain; for most Russians, such luxuries are beyond limited budgets

      b. **Attracting Foreign Investment:** most countries within the region are attracting some foreign investment; overall, connections with global economy vary, with Ukraine attracting significant capital, Moldova and Armenia making some progress

      c. **Globalization and Russia's Petroleum Economy:** Russia's oil and gas industry remains one of strongest links between region and global economy; geography of oil exports is being transformed with new pipelines; state-controlled Russian companies play large role

      d. **Local Impacts of Globalization:** local impacts highly selective, with different locations affected in distinctive ways; port cities well-positioned; locations with long history of heavy industry less positively impacted

Summary

- Huge environmental challenges remain; most are the legacy of Soviet era policies
- Declining and aging populations are continuing challenges
- Region's underlying cultural geography formed centuries ago; global influences are having local effects
- Much of region's political legacy rooted in Russian Empire and, later, Soviet Union; geopolitical relationships being redefined
- Region's economic future, especially that of Russia, remains tied to unpredictable global energy economy

Research or Term Paper Ideas

- Learn more about the on-going political tension between Russia and Chechnya. What is the source of the conflict? How and where has this conflict occurred? What have been the costs of this conflict?

- Learn more about the long-term effects of the nuclear disaster at Chernobyl. Consider the effects on people, animals, and plants. How has agriculture been impacted? What action has been taken to clean up the area? What are the long-term prospects for a recovery of the region?

- Select one of the six countries that comprise the Russian domain. Examine the continuing evolution of democratization. Is the process going smoothly? What problems have been encountered? What are the prospects for the future?

- Conduct research on the environmental degradation of Russian's northern forests. Contrast the environmental problems of Russia's forests with those of the Amazon. What are the differences? What are the similarities?

Practice Quizzes

Answers appear at the end of this book

Vocabulary Matching: Match the term to its definition.

A. Autonomous areas
B. Baikal-Amur Mainline
C. Bolsheviks
D. Centralized economic planning
E. Commonwealth of Independent States (CIS)
F. Cossacks
G. Dacha
H. Eastern Orthodox

I. Exclave
J. Glasnost
K. Gulag Archipelago
L. Perestroika
M. Permafrost
N. Russification
O. Taiga
P. Trans-Siberian Railroad

1. _____   Key central Siberian railroad connection completed in the Soviet era (1984).

2. _____   A key railroad corridor to the Pacific Ocean competed under the Russian Empire in 1904; it links Russia with the Russian Far East terminus of Vladivostok.

3. _____   A cold-climate condition of unstable, seasonally frozen ground that limits the growth of vegetation and makes construction difficult.

4. _____   A vast collection of political prisons and labor camps in which inmates often disappeared; they were located in Siberia and made famous by writer Aleksandr Solzhenitsyn.

5. _____   A form of Christianity; a loose confederation of self-governing churches in Eastern Europe and Russia that are historically linked to Byzantine traditions and to the primacy of the patriarch of Constantinople (Istanbul).

6. _____   Highly mobile, Slavic-speaking Christians of the southern Russian steppe who were known for their horsemanship skills and were pivotal in expanding Russian influence in sixteenth- and seventeenth-century Siberia.

7. _____   Vegetation type characterized by acidic soils, high occurrence of permafrost, and slow-growing fir trees which are generally small.

8. _____   A portion of a country's territory that lies outside of its contiguous land area.

9. _____   Policy that called for greater openness and political participation in the last years of the former Soviet Union.

10. _____   Economic system in which the state sets production targets and controls the means of production.

11. _____   Minor political subunits created in the former Soviet Union and designed to recognize the special status of minority groups within existing republics.

12. _____   The Soviet policy of resettling Russians into non-Russian portions of the former U.S.S.R. in order to increase Russian dominance in outlying portions of the former U.S.S.R..

13. ____ A faction of Russian communism representing interests of industrial workers, who established the former Soviet Union.

14. ____ A loose political union of former Soviet Republics (without the Baltic states) established in 1992 after the dissolution of the Soviet Union.

15. ____ Planned economic restructuring in the former Soviet Union, an early move toward a freer market.

Multiple Choice: *Choose the word or phrase that best answers the question.*

1. In addition to Russia, which other countries are parts of the Russian Domain?
   a. Ukraine and Belarus
   b. Georgia and Armenia
   c. Moldova
   d. A and B above
   e. A, B, and C above

2. Which part of the Russian domain shows the greatest damage from radioactive contamination?
   a. Ukraine and Belarus
   b. Lake Baikal
   c. Siberia and Russia
   d. Armenia and Georgia
   e. Cities along the path of the Trans-Siberian Railroad

3. What is the key to understanding the basic geographies of climate, vegetation, and agriculture in the Russian domain?
   a. The region's altitude
   b. The region's northern latitude
   c. The region's topography
   d. The region's waterways
   e. The region's seismic activity

4. What is the significance of the Ural Mountains in the Russian Domain?
   a. They rival the Himalayas in height
   b. They are the oldest mountains in the world
   c. They are the site of major development in the region
   d. They are the widest mountains in Eurasia
   e. They separate European Russia from Asian Russia

5. What is the major reason for the strikingly higher population densities in the European West part of the Russian Domain?
   a. The nearness of this region to Europe
   b. The people of this region have a culture that encourages large families
   c. The Soviet Union established this pattern with its settlement policies
   d. This is where the best agricultural lands in the region are found
   e. This region attracts many immigrants from outside the Russian Domain

6. Which of the following statements about Siberia are true?
    a. Subarctic climates are found in Siberia
    b. Populating Siberia had a political motive
    c. The Gulag Archipelago, a vast collection of political prisons, was located in Siberia
    d. A and B above
    e. A, B, and C above

7. People from which neighboring country are crossing the Amur River border (sometimes illegally) into Russia to trade, work, and live?
    a. China
    b. Finland
    c. Kazakhstan
    d. Lithuania
    e. Mongolia

8. What has been the major factor in recent migration in the Russian Domain?
    a. Deteriorating environmental conditions in some parts of Russia
    b. Extensive foreign direct investment in the Donetsk area, resulting in a booming economy there
    c. Freedom of mobility arising from the breakup of the former U.S.S.R.
    d. Internal ethnic conflicts have created many internally displaced persons
    e. Russian policies assigning people to specific cities for work and residency

9. What is the status of the population in Russia?
    a. After a short decline immediately following the breakup of the U.S.S.R., the population is now growing rapidly
    b. The population has seen a steady increase since the breakup of the U.S.S.R.
    c. The population is holding steady
    d. The population is in a state of persistent decline
    e. There are no recent population statistics available to answer this question

10. What is the dominant language group in the Russian Domain?
    a. Altaic
    b. Romance
    c. Finno-Ugric
    d. Slavic
    e. Germanic

11. The Soviet Union's invasion of this country in 1978, and the long war that followed, helped to bring about the end of the Soviet era.
    a. China
    b. Afghanistan
    c. Mongolia
    d. Poland
    e. Turkey

12.   What is the status of the Commonwealth of Independent States (CIS)?
    a.   It has become an important international trade organization
    b.   It has disbanded
    c.   It continues to exist, but it has no real economic or political power
    d.   It is actively following the path established by the European Union
    e.   It is growing in importance as a body to handle political and ethnic conflicts with the region

13.   What region of Russia tried to secure its independence from Russia after the breakup of the U.S.S.R., only to have Russia move in a large number of troops to reassert its control?
    a.   Chechnya
    b.   Siberia
    c.   Mordvinia
    d.   Sakha
    e.   Tatarstan

14.   Which of the following is an undisputed economic advantage of the Russian Domain?
    a.   Vast size and abundant natural resources
    b.   A well-educated population
    c.   Relatively high rate of urbanization
    d.   A and B above
    e.   A, B, and C above

15.   What provides the strongest global link between the Russian Domain and the rest of the world?
    a.   The region's agricultural exports
    b.   The region's cultural strengths
    c.   The region's growing software industry
    d.   The region's manufacturing sector
    e.   The region's oil and gas industry

Summary of The Russian Domain

Total Population of The Russian Domain: About 200 million

Population Indicators for The Russian Domain

	Highest (country and value)	Region Average	Lowest (country and value)
Population 2010 (millions)	Russia: 141.9	34.85	Armenia: 3.1
Density per sq km	Moldova: 122	71	Russia: 8
RNI	Armenia: 0.6	0.0	Ukraine: -0.4
TFR	Armenia; Georgia: 1.7	1.5	Moldova: 1.3
Percent Urban	Belarus: 74%	62%	Moldova: 41%
Percent < 15	Armenia: 20%	16%	Ukraine: 14%
Percent > 65	Ukraine: 16%	13%	Armenia; Moldova: 10%
Net Migration (per 1000; 2000-05)	Russia: 0.4	-2.7	Georgia: -11.5

Development Indicators for The Russian Domain

	Highest (country and value)	Region Average	Lowest (country and value)
GNI per capita/PPP 2008	Russia: $10,640	$8,210	Moldova: $2,150
GDP Avg. Annual Growth (2000-2008)	Armenia: 12.4%	8.2%	Russia: 6.2%
Human Development Index (2007)	Belarus: 0.826	0.789	Moldova: 0.720
Percent Living below $2/day	Georgia: 30.4%	10.5%	Belarus; Russia; Ukraine: <2%
Life Expectancy 2010	Armenia: 72	70	Russia; Ukraine: 68
< 5 Mortality 2008	Georgia: 30 per 1000	23 per 1000	Belarus; Russia: 13 per 1000
Gender Equity 2008	Armenia: 104	100	Georgia: 96

Chapter 10

Central Asia

Learning Objectives

➢ This chapter introduces Central Asia, which includes Mongolia, Afghanistan, and six former republics of the Soviet Union—Azerbaijan, Kazakhstan, Kyrgyzstan, Uzbekistan, Tajikistan, and Turkmenistan

➢ You will be introduced to this region's historically strategic location

➢ You will learn further about the effect of continentality and terrain on human settlement patterns

➢ The following concepts and models will be developed: desiccation, Buddhism

Chapter Outline

1. **Introduction**
    A. Central Asia encompasses a larger area than the United States; is lightly populated; and is composed of high mountains, barren deserts, and semiarid steppes (grasslands)
        a. Region is defined in various ways; most authorities agree on inclusion of five newly independent former-Soviet republics: Kazakhstan, Kyrgyzstan, Uzbekistan, Tajikistan, and Turkmenistan; this text also includes Mongolia, Afghanistan, Azerbaijan, and autonomous regions of western China (Tiber, Xinjiang, and Inner Mongolia—Nei Mongol); justification for inclusion is region's historical unity (most of region is Muslim in orientation) and environmental circumstances
        b. Historically, region has been poorly integrated into international trade networks and has been controlled largely by outside powers
2. **Environmental Geography:** relatively clean environment because of low population density; industrial pollution growing problem in larger cities; typical environmental problems of arid environments: desertification (spread of deserts), salinization (accumulation of salt in the soil), and desiccation (drying up of lakes and wetlands)
    A. **The Shrinking Aral Sea:** Until the twentieth century, the Aral Sea was the world's fourth largest lake; only significant sources of water are Amu Darya and Syr Darya Rivers; after 1950, increased diversion of rivers for irrigation; by 1970s, shoreline of lake had receded more than 40 miles; lake divided into two lakes; salinity levels increased, fish species died off
        a. Destruction of Aral Sea resulted in economic, cultural, and ecological damage; crop yields declined; desertification has accelerated; public health declined
        b. Restoration: in 2001, oil-rich Kazakhstan worked to save its portion of Aral Sea; by 2010, much of northern lake was showing signs of improvement; Uzbekistan too poor to save southern Aral Sea
    B. **Fluctuating Lakes:** World's largest lake is Caspian Sea; 15th largest is Lake Balkhash; neither Caspian Sea nor Aral Sea are seas by definition, but instead lakes—not connected with ocean
        a. Since these lakes are not drained by rivers, all naturally fluctuate depending on precipitation; but many have suffered from reduced water flow—and thus exhibit increased salinity
            i. Caspian Sea receives most water from Ural and Volga Rivers; extensive diversion for irrigation has led to decline in volume of Caspian, which undermined fisheries and caviar industry; water level began to rebound in 1980s, causing problems; current threat is pollution from oil industry
    C. **Desertification and Deforestation:** in eastern region, Gobi and Taklamakan Deserts have spread; Chinese government has attempted to stabilize with mass tree- and grass-planting

campaigns; former Soviet Central Asia also experiencing desertification; elsewhere salinization a major problem

a. Deforestation has harmed parts of region

D. **Central Asia's Physical Regions:** dominated by grassland plains (steppes) in the north, desert basins in southwest and central part, and high plateaus and mountains in south-center and southeast

    a. **Central Asian Highlands:** resulted from tectonic collision of Indian subcontinent and Asian mainland; formation of Himalayas; Karakoram Range and Pamir Mountains; other mountains radiate out from Pamir Knot, including the Hindu Kush (to southwest); the Kunlan Shan (to the east); the Tien Shan (to northwest)

        i. Tibetan Plateau; most large rivers of South, Southeast, and East Asia originate here, including the Indus, Ganges, Brahmaputra, Mekong, Yangtze, and Huang He

    b. **The Plains and Basins:** lower-lying zone divided into two main areas: central belt of deserts punctuated by lush river valleys; and northern swath of semiarid steppe

        i. **Desert belt:** divided into two segments by Tien Shan and Pamir Mountains—in the west, the arid plains of the Caspian Sea and Aral Sea basins; in the east, the Taklamakan Desert (in the Tarim Basin) and the Gobi (border between Mongolia and Inner Mongolia); western deserts contain larger rivers because of greater snowfall in Pamir Mountains

        ii. North of desert zone are expansive grasslands (steppe); then transition to Siberian taiga (coniferous forest)

E. **Global Warming in Central Asia:** because region depends on snow-fed rivers, expected to be adversely affected; 80 percent of Tibet's glaciers are retreating—thus jeopardizing dependable flows of water for local rivers; long-term implications: reduction of freshwater resources and possibility of prolonged droughts

3. **Population and Settlement:** most of region is sparsely populated; vast expanses inhabited; other areas populated by nomadic pastoralists; region is well endowed with perennial rivers and productive oases

A. **Highland Population and Subsistence Patterns**: throughout much of harsh environment, nomadic pastoralism main way of life; farming possible only in few locations; main zone of agricultural settlement is in far south, in protected valleys

    a. **Population Densities:** low throughout much of region; higher in some valleys

    b. **Mountains:** important for local people; practice of *transhumance* (moving of herds from lowland pastures in winter to highland meadows in summer); also wood supplies and freshwater

B. **Lowland Population and Subsistence Patterns:** most inhabitants live in narrow belt dividing mountains and basins and plains; population distribution forms ring around China's Tarim Basin; population west of Pamir Range concentrated in transition zone between highlands and plains along alluvial fans—fan-shaped deposits of sediments by streams flowing out of mountains (areas devoted to intensive cultivation); loess (silted soil deposited by wind) also used for agriculture

    a. **Key Valleys:** Fergana Valley of upper Syr Darya River and Azerbaijan's Kura River both support intensive agriculture and concentrated settlement

    b. **Gobi:** few sources of permanent water; scarcity of exotic rivers—thus one of region's least-populated areas

    c. **Steppes of northern Central Asia:** dominated by nomadic pastoralism; some forced to adopt sedentary lifestyles by Chinese government as means of control

C. **Population Issues:** some portions of region growing at moderately rapid pace; Afghanistan, least-developed and most male-dominated, has highest birthrate; Tajikistan also has high birthrate

    a. **Migration:** China has been resettling Han Chinese into eastern Central Asia; Russia's economic boom spurred regional migration from poorer parts of Central Asia

D. **Urbanization in Central Asia:** river valleys and oases have been partially urbanized for millennia; some early cities associated with trans-Eurasian silk route; several large cities created by Russian or Soviet planners—Tashkent (in Uzbekistan), Almaty (in Kazakhstan), and Baku (in Azerbaijan)

a. Urbanization is unevenly spreading across region

4. **Cultural Coherence:** western half of region is largely Muslim; northeastern (Mongolia) and southeastern (Tibet) Central Asia are characterized by a distinctive form of Tibetan Buddhism.

   A. **Historical Overview:** river valleys and oases of region were early sites of agricultural communities; evidence of farming villages dating to 8000 BCE in Amu Darya and Syr Darya valleys and rim of Tarim Basin; domestication of horse around 4000 BCE led to emergence of nomadic pastoralism in steppe belt

      a. **Linguistic Geography:** in first millennium CE, inhabitants spoke Indo-European languages, closely related to Persian; later replaced throughout steppe by Altaic family (including Mongolian and Turkic); southwestern Central Asia remains meeting ground of Persian and Turkic languages

      b. **Tibet:** emerged as strong, unified kingdom around 700 CE; incorporated briefly by Mongol Empire (thirteenth century)

   B. **Contemporary Linguistic and Ethnic Geography:** most of region dominated by Altaic language family speakers; some indigenous Indo-European languages remain, mostly in southwest; Tibetan remains main language of plateau; Russian also spoken in parts; Mandarin Chinese increasingly important in east—these latter two trends reflect recent population movements

      a. **Tibetan:** language divided into number of distinct dialects; usually placed in Sino-Tibetan family; relationship between Chinese and Tibetan subject to debate

      b. **Mongolian:** composed of many closely related dialects; standard Mongolian (called Khalkha) in both Mongolia and Inner Mongolia; Mongolian has own distinctive script but Mongolia adopted Cyrillic alphabet in 1941

      c. **Turkic Languages:** Turkic linguistic sphere extends from Azerbaijan in west through China's Xinjiang province in east; six main Turkic languages—five associated with newly independent republics (sixth, Uyghur, is main indigenous language of northwest China)

         i. Of six countries of former Soviet Central Asia, five are named after Turkic languages of dominant populations: Kazakhstan, Uzbekistan, Turkmenistan, Kyrgyzstan, and Azerbaijan

         ii. **Linguistic Complexity of Tajikistan:** dominated by speakers of Indo-European language; Tajik is closely related to Persian

         iii. **Language and Ethnicity in Afghanistan:** complex; region never colonized by outside power; current borders reflect pre-modern, indigenous kingdom; about half of population speaks Dari, a version of Farsi; Pashto and Dari are official languages; others also spoken

   C. **Geography of Religion:** major overland trading routes of pre-modern Eurasian crossed region; exposed to several variants of Buddhism and Islam; also Christianity and Judaism; more recently, religious lines have hardened—Islam dominates in west and center; Tibetan Buddhism in Tibet and Mongolia

      a. **Islam in Central Asia:** different interpretations of Islamic Orthodoxy; most Muslims in region are Sunnis, but Shiism is dominate in central Afghanistan, parts of Azerbaijan, and parts of Tajikistan

         i. **Chinese and Soviet Rule:** freedom of religion (both Islam and Buddhism) discouraged; both experiencing revival; in Afghanistan, the Taliban promote radical Islamic fundamentalism

      b. **Tibetan Buddhism:** dominant in Mongolia and Tibet; Buddhists in Tibet suffered persecution by Chinese from 1959 onwards

   D. **Central Asian Culture in International Context:** continuing importance of Russia; overall region is remote and poorly integrated into global cultural circuits—but there are some indicators, including increased usage of English

5. **Geopolitical Framework:** for past 300 years, region has played minor role in global political affairs

    A. **Partitioning the Steppes:** Manchu (Qing) Dynasty dominated much of region from 1644 to 1912; Russia likewise influential from 1700s onward—especially in Amu Darya and Syr Darya valleys

    B. **Central Asia Under Communist Rule:** western Central Asia dominated by Soviet Union after 1917; Tibet, Mongolia and other parts of eastern Central Asia dominated by Chinese after 1949

        a. **Soviet Central Asia:** six "republics" of western Central Asia formed part of Soviet Union—Kazakhstan, Kyrgyzstan, Tajikistan, Uzbekistan, Turkmenistan, and Azerbaijan; national identities, vague in pre-Soviet era, now given political significance during this period

        b. **Chinese Geopolitical Order:** China following communist revolution in 1949 begins to occupy Tibet in 1950; established "autonomous" regions in Xinjiang, Tiber (called Xizang in Chinese) and Inner Mongolia—autonomy largely fictitious

    C. **Current Geopolitical Tension:** region suffers from ethnic and religious tensions

        a. **Independence in Former Soviet Lands:** after 1991, six newly independent countries need to chart own political future; many still cooperated with Russia on security issues; some countries, including Tajikistan and Azerbaijan, experienced conflict and violence; others relatively easy transitions (Uzbekistan and Turkmenistan)

        b. **Strife in Western China:** continued opposition to Chinese rule; China's control of Xinjiang more secure than hold over Tibet—Xinjiang is economically more vital to China because of mineral deposits (including oil) and site of nuclear weapons tests

        c. **War in Afghanistan:** long history of conflict, including Soviet invasion and partial occupation from 1979 to 1989; dominance of fundamentalism Taliban from mid-1990s; United States invasion and war from 2001 onwards; newly installed government remains unstable and corrupt

    D. **Global Dimensions of Central Asian Geopolitics:** region emerged as key arena of geopolitical tensions—China, Russia, Pakistan, India, Iran, Turkey, and U.S. all vying for influence in region

        a. **A Continuing U.S. Role:** U.S. established military presence after 2001 in Afghanistan, Uzbekistan, Tajikistan, and Kyrgyzstan; Mongolia seeking closer ties with U.S. to counter the influence of China or Russia

        b. **Relations with China and Russia:** Russia continues to regard former Soviet republics as lying within its sphere of influence; has retained military bases in Tajikistan and Kyrgyzstan; economics and infrastructure also bind region; in 21ˢᵗ century, establishment of *Shanghai Cooperation Organization* (SCO, or Shanghai Six), composed of China, Russia, Kazakhstan, Kyrgyzstan, Tajikistan, and Uzbekistan—seeks cooperation on security issues, aims to enhance trade, serves as counterbalance to United States

        c. **Roles of Iran, Pakistan, and Turkey:** Iran major trading partner with region and offers good potential route to ocean—also has strong cultural links with region; Pakistan seeks to gain influence, hoping oil pipelines will carry oil and natural gas to its deep-water port at Gwadar; Turkey's connections are cultural—economic relations may strengthen if Central Asia orients to Western economies

6. **Economic and Social Development:** one of poorest regions of world; some countries have experienced relatively high levels of health and education—legacies of communist regimes; region plagued by inefficient economic systems; region expanding economically because of natural resources (oil and natural gas)

    A. **Economic Development in Central Asia:** increased disparity between rich and poor

        a. **Post-Communist Economies:** range of economic levels; Kazakhstan is most developed, based mostly on world's largest underutilized deposits of oil and natural gas—the Tengiz and Kashagan fields in northeastern Caspian Basin; Uzbekistan second largest economy—retained elements of command economy, world's second largest exporter of cotton, significant gold and natural gas deposits, but environmental degradation threatens cotton production; both Kyrgyzstan and Turkmenistan dependent on agriculture; Azerbaijan receiving international investment but oil-fueled economy remains poor; Tajikistan is poorest and least developed

    b. **Mongolia:** major mining investments by Chinese and Western firms resulted in rapid economic expansion

    c. **Economy of Western China:** Tibet remains burdened by poverty, though China has invested in infrastructure—and tourism is increasing; Xinjiang has economic potential because of mineral wealth

    d. **Economic Misery in Afghanistan:** Central Asia's poorest country; plagued by war and strife; significant legal exports include animal products, handwoven carpets, some fruits, nuts, gemstones; major economic activity tied to production of illicit drugs for global market

    e. **Central Asian Economies in Global Context:** Afghanistan tied because of drug trade; in former Soviet area, countries connected mostly through Russia; United States and other Western countries, and India, attracted to natural resources

B. **Social Development in Central Asia:** social conditions vary widely across region; Afghanistan ranks at bottom of every measure

    a. **Social Conditions and Status of Women in Afghanistan:** life expectancy, infant and child mortality among world's lowest; illiteracy is commonplace and gender-biased; women lead highly constrained lives; restrictions worse under conservative Taliban rule

    b. **Social Conditions in Former Soviet Republics and Mongolia:** position of women declined in many areas; women trafficked into prostitution, especially from Uzbekistan and Tajikistan; overall social welfare better in former Soviet republics, but health and educational facilities have declined with economic and political turmoil

    c. **Social Conditions in Western China:** indigenous peoples fare less well.

Summary

- Central Asia has "reappeared" on map; environmental problems brought region to world's attention
- Movements of people within region have attracted global attention—especially in Tibet
- Religious tension is a major cultural issue in region; radical Islamic fundamentalism is potent political force
- While China maintains firm grip on Tibet and Xinjiang, rest of region is key area of geopolitical competition
- Economies of region are opening up to global connections; but most still confront economic difficulties ahead

Research or Term Paper Ideas

- Conduct research on the Taliban. How did this group come to power? What are their policies? How have they remained a potent political force in the region?

- Learn more about the decline, and apparent recovery, of the Aral Sea. What is the broader impact of this environmental catastrophe? What is the long-term prognosis for this body of water?

- Contrast the movement for independence in Tibet with that of Xinjiang. How do these two movements differ? Why has the plight of Tibet received more attention in the West than the plight of Xinjiang?

- The six former Soviet republics have emerged as key sites of geopolitical competition. Learn more about the specific geopolitical motivations of Russia, China, the United States, India, and Pakistan. How do the foreign policies of these countries differ? How are they similar?

Practice Quizzes

Answers appear at the end of this book

Vocabulary Matching: Match the term to its definition.

A. Alluvial fans
B. Bodhisattva
C. Dalai Lama
D. Exclave
E. Exotic river
F. Loess

G. Pastoralists
H. Shanghai Cooperation Organization
I. Steppe
J. Taliban
K. Theocracy
L. Transhumance

1. _____   A harsh, Islamic fundamentalist political group that ruled most of Afghanistan in the late 1900s.

2. _____   A political state led by religious authorities.

3. _____   A fine, wind-deposited sediment that makes fertile soil but is very vulnerable to water erosion.

4. _____   A spiritual being who helps others attain enlightenment.

5. _____   Nomadic and sedentary peoples who rely on livestock (especially cattle, camels, sheep, and goats) for their sustenance and livelihood.

6. _____   Fan-shaped deposits of sediments dropped by streams flowing out of the mountains.

7. _____   A form of pastoralism in which animals are taken to high-altitude pastures during the summer months and returned to low-altitude pastures during the winter.

8. _____   A river that issues from a humid area and flows into a dry area which otherwise lacks streams.

9. _____   Semiarid grasslands found in many parts of the world; grasses in these areas are usually shorter and less dense than in prairies.

10. _____   A portion of a country's territory that lies outside of its contiguous land area.

11. _____   A group of countries (China, Russia, Kazakhstan, Kyrgyzstan, Tajikistan, and Uzbekistan) that seeks cooperation on such security issues as terrorism and separatism, aims to enhance trade, and serves as a counterbalance against the U.S.

12. _____   Considered to be a reincarnation of the Bodhisattva of Compassion (a spiritual being who helps others attain enlightenment).

Multiple Choice: *Choose the word or phrase that best answers the question.*

1.   What effect has Central Asia's low population density had on the region's environment?
    a.   In spite of the low population density, the fragility of the environment has caused it to become highly polluted.
    b.   Low population density has had no effect on the environment.
    c.   Low population density has helped keep the environment relatively clean.
    d.   Low population density has made it difficult for the countries of Central Asia to persuade Russia to take actions to reduce acid rain coming from Russia into the region.
    e.   The many old, polluting factories in the region have negated the effects of low population density on the region, leading to a devastate environment.

2.   Why was the Aral Sea particularly sensitive to irrigation development, resulting in an enormous loss of the lake's surface area?
    a.   It began as a small lake (by volume)
    b.   It contains freshwater, making it a direct source of water for many users
    c.   It is located in the center of a major population center, with several major cities on its banks
    d.   It is relatively shallow and its tributaries flow across a vast agricultural landscape
    e.   All of the above

3.   Which of these statements about the physical geography of Central Asia is/are correct?
    a.   Most of Central Asia is arid and it includes a central belt of deserts.
    b.   Most of Central Asia is characterized by plains and basins of low and intermediate elevation.
    c.   The mountains of Central Asia are higher and more extensive than those found anywhere else in the world.
    d.   A and C above
    e.   A, B, and C above

4.   How do most people of Central Asia make a living?
    a.   As computer software developers
    b.   As laborers in the region's factories
    c.   As pastoralists
    d.   As traditional hunters and gatherers
    e.   As workers on cotton plantations

5.   Conquest of Central Asia by which neighboring countries fostered the establishment of cities in the region?
    a.   China and Russia
    b.   India and Bangladesh
    c.   Iran and Pakistan
    d.   Nepal and Bhutan
    e.   Turkey and Georgia

6.   What is the dominant linguistic group in Central Asia?
    a.   Chinese
    b.   Indo-European
    c.   Mongolian
    d.   Tibetan
    e.   Turkic

7. Which country of Central Asia is essentially a theocracy?
    a. Kazakhstan
    b. Mongolia
    c. Tibet
    d. Uzbekistan
    e. Afghanistan

8. Which foreign language was common in Central Asia during the Soviet era?
    a. Cantonese
    b. English
    c. German
    d. Mandarin Chinese
    e. Russian

9. Which two countries of Central Asia have not been under direct control by either the Russians or the Chinese?
    a. Armenia and Kyrgyzstan
    b. Kazakhstan and Azerbaijan
    c. Mongolia and Afghanistan
    d. Tibet and Uzbekistan
    e. Turkmenistan and Tajikistan

10. What was the primary means by which early Soviet leaders (like Vladimir Lenin) tried to protect non-Russian peoples in Central Asia from Russian domination?
    a. Creation of "union republics" with some autonomy in Central Asia
    b. Enactment of laws promoting local languages and voting rights
    c. Establishment of local militias in the Central Asia republics to repel the Russian military
    d. Passage of laws forbidding encroachment by Russians into Central Asia
    e. All of the above

11. At what point did China reclaim most of its Central Asian territories?
    a. After China re-emerged as a united country in 1949
    b. After the breakup of the Soviet Union around 1990
    c. At the end of the Vietnam War
    d. At the end of World War I
    e. During the Q'ing Dynasty

12. Beginning in 1978, which country of Central Asia has experienced war with Russia, followed by internal conflict, then U.S. occupation, which continues today.
    a. Afghanistan
    b. Azerbaijan
    c. Mongolia
    d. Tibet
    e. Uzbekistan

13.   Based on conventional measures, how would one rank Central Asia in terms of its economic development?
        a.   In the middle income group of countries
        b.   Moderate economic development
        c.   One of the least prosperous in the world
        d.   One of the most prosperous regions in the world
        e.   Data are not available to make such assessments

14.   What is the most prosperous country in Central Asia?
        a.   Afghanistan
        b.   Kazakhstan
        c.   Mongolia
        d.   Tajikistan
        e.   Uzbekistan

15.   What economic activity will most likely link Central Asia to the global economy in the future?
        a.   Agricultural products
        b.   Assembly plants
        c.   Foreign direct investment
        d.   Oil and natural gas
        e.   Software programming

Summary of Central Asia

Total Population of Central Asia: 542 million

Population Indicators for Central Asia

	Highest (country and value)	Region Average	Lowest (country and value)
Population 2010 (millions)	Afghanistan: 29.1	12.9	Mongolia: 2.8
Density per sq km	Azerbaijan: 104	39	Mongolia: 2
RNI	Tajikistan: 2.4	1.7	Azerbaijan: 1.1
TFR	Afghanistan: 5.7	3.1	Azerbaijan: 2.2
Percent Urban	Mongolia: 61%	42%	Afghanistan: 22%
Percent < 15	Afghanistan: 44%	32%	Azerbaijan: 23%
Percent > 65	Kazakhstan: 8%	5%	Afghanistan: 2%
Net Migration (per 1000;2000-05)	Afghanistan: 7.5	-2.37	Tajikistan: -5.9

Development Indicators for Central Asia

	Highest (country and value)	Region Average	Lowest (country and value)
GNI per capita/PPP 2008	Kazakhstan: $9,710	$4,355	Afghanistan: $1,110
GDP Avg. Annual Growth (2000-2008)	Azerbaijan: 18.1%	10.2%	Kyrgyzstan: 4.4%
Human Development Index (2007)	Kazakhstan: 0.804	0.690	Afghanistan: 0.352
Percent Living below $2/day	Turkmenistan: 49.6%	23.6%	Azerbaijan; Kazakhstan: <2%
Life Expectancy 2010	Kazakhstan: 69	65	Afghanistan: 44
< 5 Mortality 2008	Afghanistan: 257 per 1000	69 per 1000	Kazakhstan: 30 per 1000
Gender Equity 2008	Mongolia: 104	92	Afghanistan: 58

Chapter 11

East Asia

Learning Objectives

➢ This chapter introduces East Asia, which includes China, Japan, South Korea, North Korea, and Taiwan

➢ You will learn about the historical evolution and unity of the region; you will also learn about the political and economic diversity that currently defines the region

➢ You will gain insight into the rapid economic growth, and political importance, of the region

➢ You will gain an understanding of the following key concepts: pollution exporting, Confucianism, central place theory, and ideographic writing

Chapter Outline

1. **Introduction:** East Asia is composed of China, Japan, South Korea, North Korea, and Taiwan; the region has been historically unified by cultural features—but second half of twentieth century reflects ideological, political, and economic divisions; since 1990, there has been greater integration within the region; East Asia is a core region of the global economy and has become a geopolitical power as well

2. **Environmental Geography:** environmental problems are particularly severe because of (1) large population size; (2) massive industrial development; and (3) physical geography

   A. **Dams, Flooding, and Soil Erosion in China:** Yangtze River (also called Chang Jiang) is third largest (by volume) in the world; historically, the main avenue of entry into the interior of China; since 1990s, the river has become focal point of environmental controversy

      a. **The Three Gorges Dam Controversy:** Three Gorges Dam, largest in world, completed in 2006; displaced more than 1.2 million people and caused significant ecological damage, but provides massive amounts of electricity for country, and theoretically controls flooding in lower reaches of Yangtze River

      b. **Flooding in Northern China:** the deforested North China Plain is plagued by drought and flood; several large-scale hydraulic engineering projects initiated over the years to control floods and allow irrigation, but flooding never completely prevented

         i. Worst floods occur along Huang He (or Yellow River); flows through Loess Plateau and carries huge sediment load; throughout the flat North China Plain, river loses velocity, sediments settle in riverbed, river level rises and floods surrounding region

      c. **Erosion on Loess Plateau:** Loess consists of fine, windblown sediment that was deposited on this upland area during last ice age; forms fertile soil, but washes away easily; Loess Plateau is one of poorest regions of China—good farmland is limited and drought is common

   B. **Other East Asian Environmental Problems:** deforestation, urban pollution, loss of wildlife

      a. **Deforestation and Desertification:** China lacks historical tradition of forest conservation (unlike Japan); after centuries of exploitation, many hill slopes are degraded; within past 50 years, China has attempted reforestation projects—many have failed; desertification also a major problem—China has responded with massive dune-stabilization schemes, including the "great green wall."

      b. **Mounting Pollution:** As China's industrial base expands, other environmental problems, including air and water pollution and toxic-waste dumping, have increased; problems are particularly acute in coastal areas

         i. Japan's environment is relatively clean because of strict environmental laws implemented in the 1970s; but Japan also engages in *pollution exporting*—because of strict environmental laws, many Japanese firms have moved dirtier factories overseas; Japan's pollution has been displaced to poorer countries

  ii. **Taiwan and South Korea:** both have large chemical, steel, and other heavy industries; they have imposed stringent environmental controls; however, they have also exported their pollution

 c. **Endangered Species:** growing number of endangered species linked to demand for ingredients in traditional Chinese medicine (e.g., deer antlers, bear gallbladders, snake blood, rhinoceros horn, and tiger penises); demand has increased with increased economic wealth; China has moved to protect some habitats and animals, most notably the panda bear; wildlife scarce throughout Korean peninsula but, ironically, protected along the demilitarized zone that separates North from South Korea

C. **Global Warming in East Asia:** because of China's rapid increases in carbon emissions, region has assumed global prominence in debates on climate change; potential effects of climate change throughout region are severe, including water shortages and diminished agricultural productivity

 a. **China:** insists that economic growth take precedence over reduction of greenhouse gas emissions; China also plans to significantly expand nuclear power, use of renewable energy sources, and promotion of reforestation projects

 b. **Other areas:** Japan, South Korea, and Taiwan all major emitters of greenhouse gases, but also have energy-efficient economies; Japan has been strong proponent of international treaties and is viewed as world leader in broad range of climate-friendly technologies

D. **East Asia's Physical Geography:** similar latitudinal range as United States, so climate patterns are parallel (i.e., China's southeast coast is similar to the United States' southeast coast); island belt of East Asia situated at intersection of three tectonic plates—thus geologically active with numerous volcanoes and earthquakes

 a. **Japan's Physical Environment:** Japan's extreme south is sub-tropical, northern islands are subarctic; most of country is temperate; climatic differences between locations facing Sea of Japan (colder, more snow) than locations facing Pacific Ocean; Japan is 85 percent mountainous; one of world's most heavily forested countries; limited areas of alluvial plains; largest lowland is Kanto Plain; other basins include Kansai and Nobi plains

 b. **Taiwan's Environment:** composed of large tilted block; central and eastern regions are mountainous; west is dominated by alluvial plain; extensive forests remain in eastern uplands

 c. **Chinese Environments:** China proper (excluding far west—covered in Chapter 10) exhibits considerable diversity; simple division into region north of Yangtze River valley and region south; each of these regions can be further sub-divided

  i. **Southern China**: rugged mountains, hills, with lowland basins (e.g., Sichuan Basin and Xi Basin)

  ii. **Northern China:** much of region flat plain; desertification major problem; significant sub-regions include Loess Plateau, North China Plain, and Manchuria (or *Dongbei*); Manchuria's peripheral uplands contain some of China's best-preserved forests and wildlife refuges

 d. **Korean Landscapes:** well-demarcated peninsula, partially separated from Manchuria by rugged mountains and rivers; lowlands of South Korea more conducive to agriculture; most mineral deposits, forests, and hydroelectric resources located in North Korea

3. **Population and Settlement:** most densely populated region of the world; lowlands of Japan, China, and Koreas among most intensely used portions of Earth; demographic growth has slowed considerably

A. **Agriculture and Settlement in Japan:** highly urbanized country; also one of the most mountainous; agricultural practices share same space as cities and suburbs

 a. **Japan's Agricultural Lands:** largely limited to coastal plains and basins; premier rice-growing districts along Sea of Japan coasts of Honshu; cooler climate crops (e.g., potatoes) grown in northern Japan

      b. **Settlement Patterns:** all significant cities located in same lowlands as agriculture; three largest metropolitan areas—Tokyo, Osaka, and Nagoya—sit in or near three largest plains

      c. **Japan's Urban-Agricultural Dilemma:** urban areas densely settled, but expansion would come at expense of farming areas; farmers are politically powerful in Japan

   B. **Agricultural Settlement in China, Korea, and Taiwan:** both Taiwan and Korea are highly urban; China is predominantly rural, but cities are growing

      a. **China's Agricultural Regions:** two main agricultural regions—north, dominated by wheat, millet, and sorghum; south, dominated by rice; both agriculture and settlement concentrated in broad lowlands; North China Plain as an *anthropogenic landscape*—a place that has been heavily transformed by human activities; Manchuria increasingly settled, produces food surplus; Loess Plateau—widespread soil erosion, but produces wheat and millet; also presence of subterranean housing throughout Loess Plateau

      b. **Patterns in Korea and Taiwan:** both are densely populated; populations concentrated in lowlands

   C. **Global Dimensions of Japanese Agriculture and Forestry:** Japan imports food globally; also imports forest resources, particularly from Southeast Asia, North America, and more recently Latin America and Africa; supports its own large population because of ability to import; also exports environmental degradation

   D. **Global Dimensions of Chinese Agriculture:** until 1990s, China was self-sufficient in food; rapid economic growth and loss of agricultural lands led to massive imports of food stuffs; China exports low-cost specialty groups, including garlic, apples, and farm-raised fish

   E. **Korean Agriculture in Global Context:** South Korea is main importer of wheat, corn, soybeans; North Korea pursued policy of self-sufficiency—but failed; relies on international aid, especially from South Korea

   F. **Urbanization in East Asia:** China has one of world's oldest urban foundations; Japan also has long urban tradition; Japan, Taiwan, and South Korea all around 80 percent urban

      a. **Chinese Cities:** traditional cities separated from countryside by defensive walls; most planned in accordance with geometrical principles; urban fabric changed during colonialism, with reorientation toward coast—Shanghai grew in importance; cities, especially Beijing, changed under communist rule

      b. **The Chinese Urban System:** fairly well-balanced; because of size, history, and socialist planning, largely conforms to *central place theory*---a theory that suggests an evenly distributed rural population will give rise to a regular hierarchy of urban places, with uniformly spaced larger cities surrounded by constellations of smaller cities; Hong Kong—former British colony—given unique status as "special administrative region"

      c. **Urban Patterns in Japan, Korea, and Taiwan:** South Korea and Taiwan noted for *urban primacy*; Japan exhibits *superconurbation*—a huge zone of coalesced metropolitan areas

         i. **Seoul:** primate city, capital, home of major governmental, economic, and cultural institutions; South Korea is promoting growth in other cities

         ii. **Taipei:** primate city of Taiwan

         iii. **Japan:** three major urban areas, located in three major lowlands; long history of urban growth

4. **Cultural Coherence and Diversity:** region shares certain historically rooted ways of life and systems of ideas; most traced to ancient Chinese civilization

   A. **Unifying Cultural Characteristics:** important unifying cultural characteristics related to religion and philosophical beliefs

      a. **Writing Systems:** East Asian civilizations developed *ideographic writing*—each symbol (called a character) represents an idea rather than sound; system traced back 3,200 years; with diffusion of Chinese civilization, writing system diffused to Japan, Korea, and Vietnam; in Korea and Vietnam, system largely replaced by alphabetic system; Japan modified system

         i. **Disadvantage of system:** difficult to learn; must memorize thousands of characters

         ii. **Advantage:** two literate people don't have to speak same language to understand

b. **Confucian Legacy:** as political philosophy, *Confucianism* emerged during sixth century BCE from writings of Confucius; predicated on proper authority, responsibility, and humanistic education; became part of meritocracy throughout dynastic or imperial China (roughly 220 BC through AD 1912)

c. **Modern Role of Confucian Ideology:** previously thought to hinder development because of conservatism; now argued to promote development because of social stability and respect for education; Chinese officials pushing a revival

B. **Religious Unity and Diversity:** dominance of Mahayana Buddhism; other practices

  a. **Mahayana Buddhism:** Buddhism originated in India in sixth century BCE in response to Hinduism; stresses escape from endless cycle of rebirths to reach union with divine cosmic principle (or nirvana); Buddhism split into various types, including Mahayana and Theravada Buddhism; Mahayana most widespread in East Asia, while Theravada Buddhism found in Southeast Asia

    i. **Mahayana Buddhism:** simplifies quest for enlightenment and entry to Nirvana; makes use of bodhisattvas (other people—monks—who refuse divine union for themselves in order to help others spiritually); nonexclusive, so practitioners can be both Buddhist and Taoist

  b. **Shinto:** religious practice closely bound to Japanese nationalism; began as animistic worship of nature spirits; refined into set of beliefs about nature and connections with human existence; prior to 1880s, closely intertwined with Buddhism, but Japanese authorities disentangled two faiths and elevated Shinto into nationalist cult

  c. **Taoism and Other Chinese Belief Systems:** Taoism (or Daoism) also rooted in nature and the pursuit of a balanced life; associated with feng shui (geomancy)—the practice of designing buildings in accordance with spiritual powers that course through local topography

  d. **Minority Religions:** Christianity present in East Asia; also long-established Muslim communities, especially in far west of China; temporary migrants into Japan have carried religions into country

  e. **Secularism in East Asia:** region is one of most secular regions; only small segment of people consider themselves deeply religious; in both communist China and North Korea, religion actively discouraged, replaced by allegiance to Marxism

C. **Linguistic and Ethnic Diversity:** Japanese and Chinese as distinct languages, although share writing system

  a. **Language and National Identity in Japan:** according to most scholars, Japanese not related to any other language; Korean also classified as own language family; some scholars suggest the two are related, and politics prevents linguistic relationship from being acknowledged; Japan is one of world's most homogenous countries; however, different groups in far north and far south

  b. **Minority Groups in Japan:** The Ainu live mostly on northern island of Hokkaido; speak different language; Japanese of Korean descent live in archipelago, as do Brazilians of Japanese ancestry; all minority groups experience significant discrimination; most discriminated are *Burakumin*—an outcast group of Japanese whose ancestors worked in "polluting" industries, such as leathercraft—these are among the poorest and least-educated people in Japan

  c. **Language and Identity in Korea:** people of Korean peninsula relatively homogenous; South Korea has strong sense of regional identity with historical roots; also existence of contemporary *Korean diaspora*

  d. **Language and Ethnicity Among the Han Chinese:** complex geography of language and ethnicity in China; much of country populated by Han Chinese—although not all Han speak same language; Mandarin Chinese spoken most in north; other languages prevalent in south, such as Cantonese (or Yue), Fujianese, and Shangaiese—these are true languages, not dialects, since they are not mutually intelligible; Hakka, in southern China, not considered to be Han; all languages of the Han, including Hakka, are closely related and belong to same Sinitic language subfamily; all Sinitic languages are tonal and monosyllabic

  e. **The Non-Han Peoples:** many remote upland regions of China proper are inhabited by non-Han; Chinese government recognized 55 ethnic groups; associated with self-governing village communities; expansion of Han, and acculturation, has led to declines in non-Han areas; some areas, such as Guangxi, have been designated as *autonomous regions*; autonomy is largely fictitious

  f. **Language and Ethnicity in Taiwan:** indigenous peoples located mostly in mountainous eastern region; languages related to Indonesian languages; following 1949 communist revolution in China, large numbers of Mandarin speakers fled to Taiwan

  D. **East Asian Cultures in Global Context:** region has long exhibited tensions between an internal orientation and tendencies toward cosmopolitanism; until mid-1800s, East Asian countries attempted to isolate themselves from rest of world

  a. **The Cosmopolitan Fringe:** Japan, South Korea, Taiwan, and Hong Kong exhibit vibrant internationalism, coexisting with strong national and local cultural identities; large numbers of students obtain education abroad; and East Asian culture diffusing to rest of world, as evidenced by film, video games, and comic-book culture; however, some groups within region are promoting more conservative, ultranationalist positions

  b. **Chinese Heartland:** southern coastal Chinese historically more international; interior of China largely inward looking

5. **Geopolitical Framework**

  A. **Evolution of China:** original core centered around North China Plain and Loess Plateau; several dynasties existed between 219 BCE and 1912; territory of Chinese expanded or contracted depending on dynasty

  a. **Manchu (Qing) Dynasty:** last Chinese dynasty (1644–1911); key geopolitical event was expansion into Central Asia

  b. **Modern Era:** Chinese Empire dissolved after 1911 because of internal disruption and foreign occupation; marked by first "opium war" with United Kingdom; opening up of China to global economy, particularly with establishment of treaty ports along southeastern coast; China divided by European powers into distinct *spheres of influence*; 1912 through 1920s, dominated by warlords (local military leaders)

  B. **The Rise of Japan:** strong unified state emerged in seventh century; heavily influenced by China; various feudal states up to seventeenth century

  a. **The Closing and Opening of Japan:** from c. 1600, Japan unified under Tokugawa Shogunate (a shogun was supreme military leader who theoretically remained under the emperor); during Tokugawa period, Japan largely isolated itself from rest of world; United States "opened" Japan in 1853, leading to downfall of Tokugawa Shogunate; led also to Meiji Restoration—a period of rapid modernization and militarization

  b. **The Japanese Empire:** Japan followed pattern of European colonialism; sought territories initially in neighboring islands, followed by Korea, Manchuria, North China Plain, and coastal southern China and eventually Southeast Asia; numerous wars fought during period, leading to Second World War; grand strategy was to construct a "Greater East Asia co-Prosperity Sphere," a resource-rich territory that was ruled exclusively by Japan

  C. **Postwar Geopolitics:** After World War Two, East Asia was arena of rivalry between the United States and the Soviet Union

  a. **Japan's Revival:** lost territory following World War II; military was restricted; required to obtain resources via trade; occupied by the United States

  b. **The Division of Korea:** Korean peninsula divided along 38th parallel; two separate regimes established—a capitalist South Korea, allied with the United States, and a communist North Korea, allied with the Soviet Union; Korean War (1950–1953) ended in stalemate; in 1990s, attempts to bring about reconciliation and reunification; efforts continually hindered, usually because of North Korean actions

c. **Division of China:** civil war in China from 1920s through 1945; after World War II, Chinese communists emerge victorious while nationalist Chinese flee to Taiwan; China still claims Taiwan is renegade territory; Taiwan demands recognition as independent government

d. **Chinese Territorial Issues:** China maintains control over Manchuria, Tibet, and Xinjiang; claims over territories, often leading to conflict, such as 1962 dispute with India over northeastern Kashmir; various islands in East and South China Seas; finally regained Hong Kong and Macao from former colonial powers (United Kingdom and Portugal, respectively)

D. **Global Dimensions of East Asian Geopolitics**

a. **North Korean Crisis:** North Korea's pursuit of nuclear weapons in 1990s provoked and continues to provoke international concern

b. **China on the Global Stage:** China has the largest army in world; nuclear capability; sophisticated missile technology—thus coming of age as major force in global politics

6. **Economic and Social Development:** region exhibits extreme differences in economic and social development; region as a whole (except North Korea) has exhibited remarkable growth

A. **Japan's Economy and Society:** Japan was a pacesetter between 1960s and 1980s; by 1990s, economic problems, but still world's third largest economic power

a. **Japan's Boom and Bust:** 1950s witness beginning of Japanese economic miracle and by 1980s was leader in many segments of global high-tech economy; real estate collapse in 1990s led to financial crisis and subsequent relocation of factories to Southeast Asia and China—economy further slumped as a result; huge government debt materialized; however, Japan remains core country of global economy

b. **Japanese Economic System:** bureaucracy maintains significant level of control over economy, especially compared to the United States; large groups of companies (keiretsu) are intertwined; system marked by business and social stability, though critics argue it reduces flexibility and lowers profits

c. **Living Standards and Social Conditions in Japan:** Housing, food, transportation and services very expensive; but unemployment is low; crime rates are low; measures of literacy and life expectancy among highest in world; social problems related to discrimination of minority groups; rural areas fare less well than urban

d. **Women in Japanese Society:** position of women fluctuates with overall economy; overall, women not sharing benefits of country's success

B. **The Newly Industrialized Countries**

a. **The Rise of South Korea:** after Korean War, country was devastated and impoverished; significant growth from 1960s onward with state-led economic policies; importance of huge Korean industrial conglomerates (chaebol); transition from manufacture of inexpensive consumer goods to heavy industry to high-tech; remains one of world's largest producers of semiconductors and ships; invests heavily in education; Korean multinational corporations relocating throughout Southeast Asia, Latin America, and even United States and Europe

b. **Contemporary Korea:** As middle class in South Korea expands, greater demands for democratic processes; political tensions latent; economy marked by periodic difficulties—but overall well positioned; attempts to integrate with North Korea economically, but largely unsuccessful

c. **Taiwan and Hong Kong:** both entities experienced rapid growth since 1960s; Taiwan promoted small to midsized family firms; Hong Kong forwarded *laissez-faire* economic system, with considerable market freedom; Hong Kong also transitioned from predominant trading center to major producer of textiles, toys, and other consumer goods, and business services, banking, telecommunications, and entertainment now especially important

  i. Both Taiwan and Hong Kong have strong overseas economic connections

C. **Chinese Development**

a. **China Under Communism:** command economy; failed attempts to dramatically increase industrial production (e.g., Great Leap Forward); various bouts of political repression (e.g., Cultural Revolution)

    b. **Toward a Postcommunist Economy:** from late 1970s, China has pursued free-market economy, following political struggle between pragmatists and dedicated communists; state continued to run heavy industries, but agriculture was increasingly privatized and export manufacturing promoted along southeastern coast

    c. **Industrial Reform:** establishment of *Special Economic Zones* (SEZs)—areas in which foreign investment was permitted and state interference was minimal; opened up mostly in coast regions; strategy was to attract foreign investment that could generate exports; income used to develop infrastructure and thus facilitate sustained economic growth; other market-oriented reforms followed; economic growth, however, has resulted in tension with United States—special concern related to China's holdings of U.S. treasury bonds

    d. **Social and Regional Differentiation:** since the 1970s, the Chinese government has promoted the growth of an economic elite sector, on assumption that this group would promote development throughout country; Chinese government also attempted (largely unsuccessfully) to limit population movement within country; economic disparities are pronounced and mirror economic strategies

    e. **The Booming Coastal Region:** most economic benefits located in southeastern coastal region; proximity to Taiwan and Hong Kong have helped region

    f. **Interior and Northern China:** less economic expansion; many state-owned heavy industries of Manchuria form a "rust belt"; most of interior has also missed initial wave of growth; beset with environmental degradation, underemployment, and outmigration

    g. **Rising Tensions:** uneven economic growth has generated feelings of discontentment; demands for greater political freedoms matched by government repression

  D. **Social Conditions in China:** China has achieved significant progress in social development, but human well-being varies considerably across regions; population, although declining, remains significant problem

  E. **Position of Women:** women historically had relatively low position; throughout East Asia, few women have achieved positions of power in either business or government

Summary

- Economic success of East Asia has been accompanied by severe environmental degradation; Japan, South Korea, and Taiwan imposed strict environmental regulations, but export pollution to other countries; China's economic growth poses most severe environmental threat
- East Asia is densely populated; population declines generate own problems
- East Asia united by deep cultural and historical bonds; globalization has largely been welcomed (excluding North Korea)
- Geopolitically, East Asia characterized by strife throughout twentieth century; mutual concern about growing military power in region, as well as nuclear arsenals
- With exception of North Korea, all of East Asia experienced rapid economic growth; but benefits within have been uneven

Research or Term Paper Ideas

- China's remarkable economic growth has not been matched by political freedoms. Conduct research into the different political and economic strategies forwarded by the Chinese government. Can economic growth continue without political liberties?

- Population remains a pressing issue for both China and Japan. For China, its large population size has been most pressing, while for Japan, an aging population has been of concern. In response, throughout the last half of the twentieth century, these countries have pursued very different approaches to their population problems. Conduct research and compare the differences in population problems for both China and Japan.

- Japan is an economically developed, but resource-poor country. Consequently it has been forced to obtain resources in other regions, notably Southeast Asia. In turn, however, Japan has been accused of engaging in 'shadow ecology' in that it degrades other country's environments in pursuit of its own needed resources. Conduct research into Japan's 'shadow ecology' and think about the implications for broader geopolitical issues

- The development of nuclear technology within North Korea continues to pose a global threat. Conduct research on the development of North Korea's nuclear program and examine the various treaties that have emerged in response to this threat.

- On Friday, March 11, 2011, a 9.0 magnitude earthquake occurred undersea just off the east coast of Japan. Shortly thereafter, a tsunami also struck the same region, causing further loss of life and damage to structures, including nuclear power plants in the region. In spite of being one of the best prepared countries in the world in terms of its precautions and preparations for earthquakes and other seismic activities, Japan's efforts were not enough. Research Japan's earthquake, including the country's location along the Pacific Ring of Fire, its preparedness, and the damage and recovery efforts.

Practice Quizzes

Answers appear at the end of this book

Vocabulary Matching: Match the term to its definition.

A. Anthropogenic landscape
B. Autonomous region
C. Central Place Theory
D. Confucianism
E. Diaspora
F. Geomancy
G. Ideographic writing
H. *Laissez-faire*

I. Marxism
J. Pollution exporting
K. Regulatory lakes
L. Samurai
M. Sediment load
N. Superconurbation
O. Tonal language
P. Urban primacy

1. _____ A writing system in which each symbol represents not a sound but rather a concept.

2. _____ The scattering of a particular group of people over a vast geographical area.

3. _____ A state in which a disproportionately large city dominates the urban system and is the center of economic, political, and cultural life.

4. _____ Language in which the same set of basic sounds may have very different meanings depending on the pitch in which they are uttered.

5. _____ An area that has been heavily transformed by human activities.

6. _____ The shipping of waste abroad for disposal, or constructing highly polluting factories abroad.

7. _____ Clay, silt, and sand suspended in the water of rivers and other flowing water.

8. _____ Reservoirs into which surplus water is diverted during periods of high water in order to reduce flooding downstream.

9. _____ The traditional Chinese and Korean practice of designing buildings in accordance with the principles of cosmic harmony and discord that supposedly course through the local topography.

10. _____ Economic system characterized by little government control.

11. _____ The philosophical system developed by a Chinese philosopher in the sixth century BCE; it has played a role in Chinese culture ever since.

12. _____ The warrior class of traditional Japan; after 1600, their military role declined as they assumed administrative positions, but their military ethos remained alive until the class was abolished in 1868.

13. _____ A theory used to explain the distribution of cities, and the relationships between different cities, based on retail marketing.

14. ____ In the context of China, provinces that have been granted a certain degree of political and cultural freedom from centralized authority, owing to the fact that they contain large numbers of non-Han people.

15. ____ The philosophy that promotes the necessity of a socialist economic system run through a central planning agency.

Multiple Choice: *Choose the word or phrase that best answers the question.*

1. Which of the following statements about China's Three Gorges Dam is true?
   a. The Three Gorges Dam is built on the Yangtze River and displaced 1.2 million people.
   b. The ecological and human rights consequences of the Dam are so negative that the World Bank withdrew its support.
   c. It helps prevent flooding and generates 3% of China's electricity.
   d. A and C above
   e. A, B, and C above

2. Which country of East Asia has a very long history of forest conservation?
   a. China
   b. Japan
   c. North Korea
   d. South Korea
   e. Taiwan

3. Where are most of Japan's agriculture, industry, and human settlements located?
   a. Coastal plains
   b. Interior basins
   c. On Japan's northern-most island
   d. A and B above
   e. A, B, and C above

4. Which of these products does Japan import in unusually high amounts?
   a. Food
   b. Energy
   c. Manufactured goods
   d. A and B above
   e. A, B, and C above

5. Only one country of East Asia is still mostly rural; which one is it?
   a. China
   b. Japan
   c. North Korea
   d. South Korea
   e. Taiwan

6. Which geographic theory best explains the distribution of Chinese cities?
    a. Central Place
    b. Concentric Zone
    c. Counterurbanization
    d. Historical Evolution of the City
    e. Urban Realms

7. What is the basis for the Chinese ideographic writing system?
    a. Syllables
    b. Phonetic sounds
    c. Ideas, symbols
    d. Hieroglyphics
    e. Greek

8. What religion is most widespread in East Asia?
    a. Animism
    b. Buddhism
    c. Christianity
    d. Hinduism
    e. Islam

9. Up until the mid-1980s, what was the approach of the countries of East Asia toward the West?
    a. They adopted Western technology
    b. They attempted to insulate themselves from Western cultural values
    c. They cautiously worked with Western countries
    d. They embraced Western culture
    e. They traveled to the West in search for education

10. The political history of East Asia revolves around the centrality of which country in the region?
    a. China
    b. Japan
    c. North Korea
    d. South Korea
    e. Taiwan

11. What country of East Asia has been a source of concern because of its development of missiles that are capable of carrying nuclear warheads?
    a. China
    b. Japan
    c. North Korea
    d. South Korea
    e. Taiwan

12. Which other country of East Asia does the government of China consider to be an integral part of China?
    a. Hong Kong
    b. Japan
    c. North Korea
    d. South Korea
    e. Taiwan

13. What model did South Korea and Taiwan follow on their path to development?
    a. The British model of colonial dominance
    b. The Chinese model of communism
    c. The German model of industrial development
    d. The Japanese model of export-led economic growth
    e. The Soviet model of collectivization

14. Which of the following pairings of East Asian countries with their approaches to economic development is accurate?
    a. China has taken a modified capitalist road to development
    b. Hong Kong has one of the most *laissez-faire* economic systems in the world
    c. South Korea initiated a program of export-led economic growth
    d. A and B above
    e. A, B, and C above

15. Which country of East Asia has had a long-term population control policy that has resulted in a gender imbalance?
    a. China
    b. Japan
    c. North Korea
    d. South Korea
    e. Taiwan

Summary of East Asia

Total Population of East Asia: About 1.56 billion

Population Indicators for East Asia

	Highest (country and value)	Region Average	Lowest (country and value)
Population 2010 (millions)	China: 1,338.1	261.2	Hong Kong: 7.0
Density per sq km	Hong Kong: 6,410	1,368	China: 140
RNI	Hong Kong: 0.6	0.4	Japan: 0.0
TFR	North Korea: 2.0	1.4	Hong Kong; Taiwan: 1.0
Percent Urban	Hong Kong: 100%	75%	China: 47%
Percent < 15	North Korea: 22%	16%	Hong Kong: 12%
Percent > 65	Japan: 23%	12%	China: 8%
Net Migration (per 1000;2000-05)	Hong Kong: 3.3	0.6	China: -0.3

Chapter 12

South Asia

Learning Objectives

- ➢ This chapter introduces the countries of South Asia, including India, Pakistan, Bangladesh, Sri Lanka, Nepal, Bhutan, and the Maldives
- ➢ You will gain a deeper understanding and appreciation of the impact of climate on agricultural patterns; this is accomplished though an in-depth discussion of the monsoon
- ➢ You will learn the challenges of feeding a large and growing population
- ➢ You will be familiar with the physical, cultural, political, and economic landscapes of South Asia
- ➢ The following concepts will be introduced and elaborated: monsoon, Green Revolution, orographic precipitation, forward capital, Hinduism, and the caste system

Chapter Outline

1. **Introduction**
   A. South Asia is a region of deep historical and cultural connections
   B. It is a region of intense political conflict—much of which is based on religion and/or ethnicity
   C. There are significant demographic and economic concerns throughout the region
   D. Although poverty is rampant, the region exhibits a significant degree of vitality in recent decades
2. **Environmental Geography:** South Asia exhibits a diversity of environments and complex environmental issues
   A. **Building the Quadrilateral Highway:** completed in 2010, this highway connects India's four largest cities (New Delhi, Kolkata, Chennai, and Mumbai); the highway also highlights many of the region's environmental issues (i.e., pollution and deforestation)
   B. **Environmental Issues in South Asia**
      a. **Natural Hazards in Bangladesh:** link between population pressure and environmental problems well-illustrated in delta area of Bangladesh; physical topography, combined with climatic factors and high population densities create a significant hazard
      b. **Forests and Deforestation:** much of South Asia, because of historical settlement patterns, has been deforested; railroad construction in the nineteenth century intensified deforestation; most recently, hill slopes in Himalayas and eastern India have been logged for commercial purposes; reforestation is occurring—but replanted areas support little wildlife
         a. Because of deforestation, many villages suffer from fuel shortages; widespread use of cattle dung for fuel—but this diverts needed nutrients from soil and contributes to air pollution
         b. Recent history of environmental activism, such as the Chipko movement in northern India
      c. **Wildlife: Extinction and Protection**
         a. Situation in region is rather bleak; population pressures and poverty contribute
         b. Preservation efforts in South Asia exceed those in other parts of Asia; Project Tiger, in India, is credited with increasing country's tiger population

    C. **Four Sub-regions of South Asia**
        a. **Mountains of the North:** dominated by Himalayan Range, former northern borders of India, Nepal, and Bhutan; mountains linked to Karakoram Range to the west and the Arakan Yoma Mountains to the east; entire region is seismically active
        b. **Indus-Ganges-Brahmaputra Lowlands:** south of mountains; lowlands created by three major river systems; region composed of vast alluvial plains of fertile and easily farmed soils
            a. **Indus River:** longest, flows southward from Himalayas through Pakistan to Arabian Sea; provides irrigation water to Pakistan's desert regions; Indus Delta is sparsely populated
            b. **Ganges River:** southeastward from Himalayas to the Bay of Bengal; provides fertile alluvial soil in northern India; vital transportation corridor; Hindus consider the Ganges to be sacred; one of world's most polluted watercourses
            c. **Brahmaputra River:** headwaters in Tibetan Plateau, joins the Ganges River in central Bangladesh; the Ganges-Brahmaputra Delta is densely settled
        c. **Peninsular India:** consists primarily of Deccan Plateau and bordered by narrow coastal plains backed by two elongated north-south trending mountain ranges (Western Ghats and Eastern Ghats); soils are poor or average on Deccan Plateau—better in state of Maharashtra; reliable water supply for agriculture is major problem; much of western region lies in rain shadow of Western Ghats; groundwater has been severely depleted; large dams have been developed—but are controversial
        d. **The Southern Islands:** consists of Sri Lanka—a tear-shaped island with extensive coastal plains and mountainous southern interior; and the Maldives—a chain of more than 1,200 islands off the southwestern tip of India; combined land area of the Maldives is about 116 square miles; only a quarter of the islands are inhabited; these are low, flat coral atolls, with maximum elevation of just over six feet above sea level
    D. **South Asia's Monsoon Climates**
        a. **Monsoon:** distinct seasonal change between wet and dry periods; South Asia has warm, rainy summer monsoon (June–October) and cold, dry winter monsoon (November–February)
        b. **Orographic Rainfall:** results from uplifting and cooling of moist winds; sheltered (or lee) side of mountains usually very dry with little precipitation; this dry area is called a *rain shadow*, and experiences a *rain shadow effect*; areas to the east of the Western Ghats experience a rain shadow effect
    E. **Global Warming in South Asia:** many areas especially vulnerable; minor rise in sea level will inundate large areas of Ganges-Brahmaputra Delta; several small islands in this area have already disappeared; some climate models predict that the entire atoll archipelago of the Maldives will disappear
        a. **Agriculture:** will be adversely affected; retreating glaciers will lead to decreased water in rivers, thus threatening irrigated areas; some areas might experience increased rainfall—leading to extensive flooding
        b. **Attitudes:** India signed Kyoto Protocol—India's growing economy, with large population size, contributes to increased greenhouse gas emissions; government leaders argue that responsibility for global climate change should come from industrialized countries
3. **Population and Settlement:** South Asia will soon surpass East Asia as most populous world region; fertility levels have declined, but population continues to grow rapidly through much of region
    A. **The Geography of Population Expansion**: decline in fertility levels shows distinct geographical patterns; most of southern and eastern India, along with Sri Lanka, should witness population stabilization; northern regions will continue to increase
        a. **India:** widespread concern over population began in 1960s; introduction of family planning programs; total fertility rate dropped—related to increased women's education and family planning; ongoing problem of preference for sons; sex-selective abortion is illegal—but happens

    b. **Pakistan and Bangladesh:** government exhibits ambivalent attitude toward family planning; official position is that birthrate is excessive—but only recently moved to lower overall rate; Bangladesh has made significant strides in reducing birthrates through family planning

B. **Migration and the Settlement Landscape:** most densely settled areas coincide with zones of fertile soils and dependable water supplies

    a. Largest rural populations found in core area of Ganges and Indus River Valleys and coastal plains of India; settlement is less dense on Deccan Plateau; major population clusters in Katmandu Valley of Nepal and the Valley of Kashmir in northern India

    b. Bangladesh and northern India are zones of out-migration—desperate conditions fuel rural-urban migration; other regions experience out-migration following conflict or war

    c. South Asia is one of the least urbanized regions in the world; most inhabitants continue to live in compact rural villages

C. **Agricultural Regions and Activities:** historically, region has been less productive than East Asia; causes complex, including colonial history, social status of farmers, and land ownership issues

    a. **Crop Zones:** South Asia divided into distinct agricultural regions

        a. **Rice:** grown mostly in lower Ganges Valley, lowlands of India's eastern and western coasts, delta region of Bangladesh, Pakistan's lower Indus Valley, and Sri Lanka

        b. **Wheat:** grown mostly in northern Indus Valley and western half of India's Ganges Valley; Punjab—the region's "breadbasket"—is located in northwestern India and parts of Pakistan; Green Revolution successful in increasing grain yields

        c. **Millet and Sorghum:** grown throughout less-fertile areas of central India; also root crops (e.g., manioc)

        d. **Other Crops:** some commercially produced, others for local subsistence; oil seeds (e.g., sesame and peanuts) grown in semi-arid districts; Sri Lanka and Indian state of Kerala produce coconuts, spices, and tea; cotton widely cultivated in Pakistan and west central India; Bangladesh major producer of jute—a tough fiber used in manufacture of rope

    b. **Green Revolution:** originated during 1960s; cross-breed crops to make disease-resistant; efforts initially successful and widely diffused; South Asia transformed from region of chronic food deficiency to one of self-sufficiency

        a. **Disadvantages:** critics argue that practice is destructive to environment; practice is expensive and creates a situation of haves and have nots

    c. **Future Food Supply:** unclear whether benefits of Green Revolution will continue; efforts to expand water delivery; problem of soil salinization remains

D. **Urban South Asia:** Region is one of world's least urbanized, but contains some of world's largest urban agglomerations; 43 cities have populations in excess of one million

    a. **Rapid growth:** has led to many problems—homelessness, poverty, congestion, water shortages, air pollution, and sewage disposal

        a. **Delhi:** contains about 19 million people; two landscapes—Old Delhi and New Delhi; the former a planned city developed since 1911; rapid growth and unregulated automobile and industrial emissions made Delhi one of world's 10 most polluted cities; significant environmental progress has been made

        b. **Kolkata:** emblematic of problems faced by rapidly growing cities; over 15 million residents—suffer from water and power shortages, and lack of adequate sewage treatment; considerable homelessness; streets routinely flood during rainy season; ethnic tensions, troubled economic base, and overload infrastructure add to problems

        c. **Dhaka:** capital and primate city of Bangladesh; experienced rapid growth from rural-urban migration; population of about 13 million; center of economic vitality, and is global center for clothing, shoe, and sports equipment

        d. **Karachi:** rapidly growing port city of 18 million in Pakistan; country's largest urban area and commercial core; former capital until planned construction of Islamabad in 1963; suffers political and ethnic tensions

  e. **Islamabad:** planned city developed in 1963 as capital of Pakistan; city is considered a *forward capital* because it signals, both symbolically and geographically, the intentions of the country to support claims of disputed northern territory of Kashmir (controlled by India)

4. **Cultural Coherence and Diversity:** historically, South Asia forms well-defined cultural region; arrival of Islam added new religious dimension; since mid-twentieth century, religious and political strife have increased; since 1980s, growth in Hindu fundamentalism, or more properly called *Hindu nationalism*—the promotion of religious values of Hinduism as essential fabric of Indian society; rising Islamic fundamentalism in Pakistan

   A. **Origins of South Asian Civilizations:** culture extends to Indus Valley more than 4,000 years ago; newer settlements around 800 BCE in middle Ganges Valley

      a. **Hindu Civilization:** Hinduism—faith that incorporates diverse forms of worship; lacks standard system of beliefs; all Hindus share common set of epic stories; Sanskrit is sacred language of Hinduism; hallmark is belief in transmigration of souls from being to being through reincarnation—nature of one's acts in physical world influences the courses of future lives; Hinduism has a caste system—strict division of society into different hierarchically ranked hereditary groups

      b. **Buddhism:** Siddhartha Gautama (the Buddha), a prince, was born in 563 BCE; he rejected life of wealth and power and sought path to enlightenment with the cosmos—argued that this path to nirvana was open to all—thus, rejection of caste system; his followers established Buddhism as new religion; in later centuries, Buddhism diffused throughout East, Southeast, and Central Asia; Buddhism never replaced Hinduism in South Asia; widely practiced in Sri Lanka and throughout Himalayas

      c. **Arrival of Islam:** significant advance with Turkish-speaking Muslims around year 1000; by 1300, most of South Asia controlled by Muslims, but Hindu kingdoms persisted in southern India; during sixteenth and seventeenth centuries, the Mughal Empire dominated much of region; Hindus from lower castes especially attracted to Islam and converted; although very different, both Hinduism and Islam coexisted peacefully until twentieth century

   B. **The Caste System:** although unifying feature of South Asia, system not uniformly distributed; in modern Pakistan and Bangladesh, role is fading; in Buddhist areas, influence is minimal; in India, caste system even deemphasized—but continues to structure day-to-day lives

      a. **Caste:** term is of Portuguese origin—combines two distinct local concepts: *varna* (ancient fourfold social hierarchy of Hindu world) and *jati* (hundreds of local endogamous groups that exist at each varna level—sub-castes)

      b. **Main Caste Groups:** three groups constitute traditional elite

         a. **Brahmins:** traditional priestly caste; traditional intellectual elite

         b. **Kshatriyas:** warrior or princely caste; usually ruled old Hindu kingdoms

         c. **Vaishyas:** traditional merchant caste

      c. **Fourth Category:** *Sudras*—composed of large array of sub-castes, originally reflecting occupational groups

      d. **Dalits**: group that stands outside varna system—the untouchables; low-status positions derived from "unclean" occupations, such as leather workers, scavengers, latrine cleaners, and swine herders; many Indian dalits converted to Islam, Christianity, and Buddhism; discrimination continues

      e. **The Changing Caste System:** system is in a state of flux; concept of "untouchability," for example, technically illegal in India—but continues

   C. **Contemporary Geographies of Religion:** region exhibits predominantly Hindu heritage with substantial Muslim imprint

      a. **Hinduism:** minority religion in Pakistan, Bangladesh, and Sri Lanka; widespread in India and Nepal

    b. **Islam:** minority religion for region as a whole; overwhelming majority in Bangladesh and Pakistan; in India, most concentrated in four areas: India's cities; Vale of Kashmir; upper and central Ganges plain; in southwestern state of Kerala

    c. **Sikhism:** originated in late 1400s in Punjab; combined elements of both Islam and Hinduism; concentrated overwhelmingly in Indian state of Punjab

    d. **Buddhism and Jainism:** Buddhism prevalent in Sri Lanka, the Himalayas; Jainism—originated around 500 BCE as protest against orthodox Hinduism; non-violent religion—agriculture is forbidden because plowing might harm insects; concentrated in northwestern India, particularly Gujarat

    e. **Other Religious Groups**: Parsis (or Zoroastrians) concentrated in Mumbai—arrived with refugees from Persia after seventh century—focuses on cosmic struggle between good and evil; many practitioners in business community; other religions including Christianity and Judaism scattered throughout region

  D. **Geographies of Language**: important dividing line between north and south; northward are Indo-European languages; southward are Dravidian languages (a family unique to South Asia); scattered Austra-Asiatic languages throughout

    a. **Indo-European North:** Iranian languages (e.g., Baluchi and Pashtun) found in western Pakistan; Indo-Aryan languages are closely related—Hindi is most widely spoken in South Asia, with Bengali second; Urdu is official language of Pakistan

    b. **Languages of the South:** four main Dravidian languages, each associated with an Indian state: Kannada in Karnataka, Malayalam in Kerala, Telugu in Andhra Pradesh, and Tamil in Tamil Nadu; Sinhalese is dominant language of Sri Lanka; Divehi (a Sinhalese dialect) is national language of the Maldives

    c. **Linguistic Dilemmas:** multilingual countries of Sri Lanka, Pakistan, and India marked by linguistic conflicts; Indian nationalists want national language; Indian government recognizes 23 official languages

    d. **Role of English:** English is widespread and becoming "associate official language" of India; English-medium schools increasing

  E. **South Asians in a Global Cultural Context:** widespread use of English has facilitated spread of global culture; South Asian culture also diffused outwards; diffusion of South Asian culture facilitated by Indian diaspora—the out-migration of peoples from India to other world regions

    a. **Problems:** globalization of culture has brought tensions; traditional Muslim and Hindu religious customs are conservative and disapprove of outward displays of sexuality—a problem with infusion of much Western culture

5. **Geopolitical Framework:** prior to the 1800s, South Asia never politically united; British rule provided unification; ongoing geopolitical tensions over borders

  A. **South Asia Before and After Independence in 1947:** during 1500s, most of northern subcontinent ruled by Mughal Empire—Muslim state ruled by people of Central Asian descent; Southern India under control of Hindu kingdom; Portuguese and Dutch began colonization

    a. **British Conquest:** British and French displaced Portuguese and Dutch; competed with each other for trading posts; Britain emerges as most dominant—led by British East India Company—private firm that acted on behalf of British government; French restricted to coastal possession in southern India

    b. **From Company Control to British Colony:** Following rebellion in 1857, British government ruled India; maintained direct control over most productive and densely settled areas; ruled Sri Lanka; administered subcontinent from three coastal cities that they largely created: Bombay (Mumbai), Madras (Chennai), and Calcutta (Kolkata)

    c. **Independence and Partition:** in early twentieth century, people of South Asia demanded independence; massive political protests followed; ethnic and religious differences posed dilemma for independence movement; after World War II, British withdrew (1947); region was divided into India and Pakistan—violence ensued, as did mass migrations; Pakistan was created as two-part country, its western section in the Indus Valley and its eastern portion in the Ganges Delta; rebellion in 1971 led to creation of independent Pakistan in west and independent Bangladesh in east; Pakistan has remained politically unstable and prone to military rule; Bangladesh has experience political problems

    d. **Geopolitical Structure of India:** since 1947, leaders committed to democracy; organized as federal state, with significant amount of power given to individual states—these constituent states reorganized to match country's linguistic geography; several new states have been added; smaller linguistic/ethnic groups still demand greater autonomy

B. **Ethnic Conflicts in South Asia**

    a. **Kashmir:** during British period, Kashmir was large state with Muslim core joined with Hindu district in south; later ruled by Hindu *maharaja*—king subject to British advisors; during partition, both India and Pakistan fought for region; two countries continue to fight inconclusive wars over region; 87 percent of people of Kashmir desire independence—though neither India nor Pakistan want that

    b. **Punjab:** religious conflict has caused political tensions; area originally composed of Hindu, Muslim, and Sikh; divided in 1947; Sikhs demand independence

    c. **Northeastern Fringe:** upland areas of India's northeast; various ethnic groups demand greater autonomy

    d. **Sri Lanka:** ongoing interethnic violence between Hindu Tamils (of north) and Buddhist Sinhalese (of the south); Sinhalese nationalists favor unitary government while Tamils demand political and cultural autonomy; since 1983, Tamil Tigers (Liberation Tigers of Tamil Eelam) continue insurgency

C. **Maoist Challenge:** Not all conflicts rooted in ethnic or religious factors—poverty and inequality have generated revolutionary movements; many of these find inspiration in former Chinese communist leader Mao Zedong; Nepalist Maoists, in particular, have emerged as significant force in 1990s—frustrated by lack of development in rural areas

D. **International Geopolitics:** major international geopolitical problem is struggle between India and Pakistan—both have nuclear capability; Pakistan allied with the United States during Cold War; India remained neutral but leaned to Soviet Union; China's military connection with Pakistan rooted in past war (1962) with India over territory in northern Kashmir; trade has lessened tension between China and India

    a. **Pakistan:** complex geopolitical situation after attacks on United States on September 11, 2001; Pakistan had supported Afghanistan's Taliban regime; U.S. forced Pakistan to either assist the U.S. against Taliban and receive debt reductions and other forms of aid, or it would lose favor; Pakistan decided to support the U.S., sparking protests within its country; Pakistan continues to face Islamist insurgency, geopolitical conflict with India, low-level ethnic rebellions in southwestern province

    b. **India's Changing Geopolitical Situation:** relations with Bangladesh deteriorated in late 1990s because of India's concerns over illegal Bangladeshi immigration and fears that Bangladesh was providing refuge to separatist fighters from India's northeast

6. **Economic and Social Development:** South Asia characterized by developmental paradoxes: one of poorest world regions, yet site of some immense fortunes; many scientific and technological accomplishments, but some of world's highest illiteracy rates; emerging as center of global information economy, but as whole was long one of world's most self-contained and inward-looking regions; poverty is widespread; many suffer from undernourishment, malnutrition—infant mortality is high; but region (especially India) has large and growing middle class and small but wealthy upper class

A. **Geographies of Economic Development:** since 1990s, countries (especially India) have opened economies to global economic system; core areas of economic development have emerged, but peripheral areas have lagged—resulting landscape of economic disparity

    a. **The Himalayan Countries:** both Nepal and Bhutan disadvantaged by rugged terrain and remote locations; remain relatively isolated from modern technology and infrastructure; many areas subsistence-oriented; Bhutan has purposely remained isolated from global economy—

government promotes "gross national happiness" instead of "gross national product"; Bhutan also exports substantial amounts of hydroelectricity to India; Nepal more heavily populated and suffers severe environmental degradation; tourism has brought some prosperity to Nepal, but also suffers; remittances from Nepalese migrants help sustain economy

b. **Bangladesh:** poverty is extreme and widespread; environmental degradation contributes to impoverishment; agricultural emphasis on jute has suffered because of world's use of synthetic materials, which undercut global jute market; country has emerged as important textile and clothing manufacturer—in part because of low wage rates

c. **Pakistan:** inherited fairly well-developed infrastructure after independence; shares fertile Punjab with India; has large textile industry because of cotton production; but since 2008, economy faltering—country burdened by military spending; unable to develop successful IT industry; has inadequate energy supplies; constructed new deep-water port at Gwadar with Chinese engineering and financial assistance, and managed by Port of Singapore

d. **Sri Lanka and the Maldives:** Sri Lanka is second most highly developed in South Asia; exports concentrated in textiles and agricultural products (rubber and tea); but remains poor; hopes to benefit from port, high levels of education, and tourism potential; the Maldives' economy is based on fishing and international tourism—both vulnerable

e. **India's Less-Developed Areas:** poorer regions located in north and east; extreme poverty in lower Ganges Valley; Bihar is India's poorest state; most areas dominated by subsistence economies

f. **India's Centers of Economic Growth:** found mostly in south and west; west-central states of Gujarat and Maharashtra noted for industrial and financial clout and agricultural productivity; benefit from connections with Indian diaspora; Mumbai, located in Maharashtra, is financial center, media capital and manufacturing powerhouse; large industrial zones located in state; center of India's fast-growing high-technology sector in the south, especially in city of Bengaluru (formerly Bangalore)—known as the "Silicon Plateau"; other high-tech centers include Hyderabad (often called "Cyberabad"); and Chennai

B. **Globalization and India's Economic Future:** not one of world's most globalized regions, but globalization is advancing rapidly; does not yet receive substantial amounts of direct foreign investment as found in China; consequently, infrastructure overall remains poorly developed and inadequate; electric supply is inadequate

C. **Social Development:** levels of social well-being vary widely across region; people in more prosperous areas are healthier, live longer, and are better educated

a. **The Educated South:** Sri Lanka considered success story of social development; government funds universal primary education and inexpensive health clinics; Kerala, in southwest India, is crowded and has high unemployment—but indices of social development are best in India—in part because Kerala led by socialist party that stressed mass education and community health care

b. **The Status of Women:** women throughout region overall have low social position; both Hindu and Muslim traditions are limited for women; some places exhibit pronounced discrimination—such as Indus-Ganges Basin; even in middle-class households, women suffer discrimination and limitations; dowry increasing in parts, as are bride-murders—laws against practice have been ignored; social bias against women less evident in southern India and Sri Lanka

Summary

- South Asia, though large, has been somewhat overshadowed by other regions; but increasingly, South Asia figures prominently in discussions of world problems and issues
- Environmental degradation and instability pose particular problems; rising sea level directly threatens the Maldives and Bangladesh
- Continuing population growth demands attention; fertility rates have declined, but countries still unable to support expanding populations
- South Asia exhibits diverse cultural heritage; ethnic and/or religious conflicts increased throughout twentieth century
- Geopolitical tensions within region remain—notably border dispute between India and Pakistan over Kashmir
- Although region remains one of poorest of the world, South Asia has seen rapid (though geographically varied) economic expansion

Research or Term Paper Ideas

- Conduct research to learn more about one of the many religious conflicts in South Asia. Who are the participants? How did the conflict arise? What strategies (political or military) have the participants pursued? What is the status of the conflict? What is the long-term prospect for a peaceful resolution of the conflict?

- Learn more about the nuclear rivalry between India and Pakistan. How did each country acquire nuclear weaponry capabilities? What is the position of these two countries vis-à-vis international nuclear non-proliferation treaties?

- Conduct research on the on-going dispute in Kashmir. How does this conflict compare with other disputed territories, such as the attempts in Southwest Asia to establish a Palestinian homeland or a Kurdish homeland? What lessons could be drawn from these different territorial disputes?

- The geopolitical position of Pakistan is very complex, with seemingly shifting alliances and objectives. Learn more about the foreign policy of Pakistan. How must Pakistan balance its foreign relations with domestic concerns?

Practice Quizzes

Answers appear at the end of this book

Vocabulary Matching: Match the term to its definition.

A. British East India Company	I. Monsoon
B. Caste system	J. Mughal Empire
C. Dalits	K. Orographic rainfall
D. Green Revolution	L. Rain-shadow effect
E. Indian diaspora	M. Salinization
F. Jainism	N. Sikhism
G. Linguistic nationalism	O. Subcontinent
H. Maharaja	P. Tamil Tigers

1. _____ Linking of a specific language with political or national goals.

2. _____ The area of low rainfall found on the leeward (downwind) side of a mountain range.

3. _____ Term applied to the development of agricultural techniques used in developing countries that usually combine new, genetically altered seeds that provide higher yields than native seeds when combined with high inputs of chemical fertilizer, irrigation, and pesticides.

4. _____ Enhanced precipitation over uplands that results from lifting (and cooling) of air masses as they are forced over mountains.

5. _____ A large segment of land separated from the main landmass on which it sits by lofty mountains or other geographical barriers; South Asia is often called the "Indian Subcontinent."

6. _____ Complex division of South Asian society into different hierarchically ranked hereditary groups, most explicit in Hindu society.

7. _____ The distinct seasonal change of wind direction, which is the dominant climatic factor for most of South Asia.

8. _____ The migration of large numbers of Indians to foreign countries.

9. _____ "Untouchables," people whose low status derived historically from their working in "unclean" occupations.

10. _____ The buildup of salt in agricultural fields.

11. _____ The preeminent Islamic period of rule that covered most of South Asia during early sixteenth to late seventeenth centuries and attempted to unify both Muslims and Hindus into a large South Asian state.

12. _____ An Indian religion combining Islamic and Hindu elements, founded in the Punjab region in the fifteenth century.

13. _____ A religion that emerged in northern India and stressed non-violence, taking this creed to its ultimate extreme.

14. _____ The rebel force in Sri Lanka, fighting against the Sinhalese majority with tensions remaining.

15. _____ Private organization that acted as an arm of the British government, and was free to stake out a South Asian empire of its own.

Multiple Choice: *Choose the word or phrase that best answers the question.*

1. What are the tallest mountains in South Asia --- and the world?
   a. Eastern Ghats
   b. Himalayas
   c. Satpura Range
   d. Vindhya Range
   e. Western Ghats

2. Why is Bangladesh so vulnerable to the cyclones and wet monsoons of South Asia?
   a. Concentration of the agricultural activity of Bangladesh in the fertile delta valley
   b. High density and clustering of the population of Bangladesh on the low-lying delta area
   c. Deforestation of the headwaters of the Ganges and Brahmaputra Rivers
   d. A and B above
   e. A, B, and C above

3. Which country of South Asia protects its wildlife better than most other countries of Asia?
   a. Bhutan
   b. Bangladesh
   c. Pakistan
   d. India
   e. Nepal

4. Which of the following statements about monsoons is/are true?
   a. Monsoons are the dominant climatic force in most of South Asia
   b. A monsoon is a distinct seasonal change of wind direction, and can be wet or dry
   c. Monsoon is the Hindi world for El Niño
   d. A and B above
   e. A and C above

5. Which of the following statements is true?
   a. South Asia is the most populous region in the world
   b. South Asia is growing at a slower rate than East Asia
   c. A cultural preference for males in South Asia complicates family planning efforts
   d. India has the highest population in the world
   e. Bangladesh has South Asia's highest TFR (total fertility rate)

6.   Which country of South Asia has made significant strides in family planning by employing a large number of women as fieldworkers who take information to other women in the country's villages?
   a.   Bangladesh
   b.   India
   c.   Nepal
   d.   Pakistan
   e.   Sri Lanka

7.   Which part of South Asia has seen its TFR (total fertility rate) drop to just below replacement level?
   a.   Bangladesh
   b.   India's state of Punjab
   c.   Pakistan
   d.   Maldives
   e.   Sri Lanka's Tamil region

8.   Where do most of the people of South Asia live?
   a.   In the far north highland
   b.   In the cities
   c.   In the arid lands of the northwest
   d.   In compact rural villages and small towns
   e.   In apartments

9.   Why do so many South Asians receive inadequate protein?
   a.   Poverty
   b.   Most Hindus are vegetarians
   c.   Meat is expensive
   d.   A and B above
   e.   A, B, and C above

10.   Which of the following pairings of place with religion in South Asia is INCORRECT?
   a.   Sri Lanka – Buddhism
   b.   Pakistan – Islam
   c.   Nepal – Jainism
   d.   India – Hinduism
   e.   Bangladesh – Islam

11.   Which of the following statements about English in India is/are true?
   a.   English is the main integrating language of India and was brought by British colonizers
   b.   Indian children are taught English early, regardless of their social and economic class
   c.   Information technology companies in India sometimes ask their employees to watch American TV shows in order to gain fluency in American pronunciation and slang
   d.   A and B above
   e.   A and C above

12. What was the main reason for partitioning South Asia into three countries (India, Pakistan, Bangladesh) after it achieved its freedom from Britain?
    a. Ethnic rivalries had erupted into violence and there was no other choice
    b. Linguistic differences made unity impossible
    c. Mountains separated these regions, so it was decided to create a physiographic boundary
    d. Muslims in the northeast and northwest parts of the region did not wish to live under Hindu rulers
    e. The boundaries were a legacy of British colonization, much as the Berlin Conference determined the borders of today's African countries

13. Which parts of South Asia has nuclear weapons?
    a. India
    b. Pakistan
    c. Bangladesh
    d. A and B above
    e. A and C above

14. Which parts of South Asia are economically disadvantaged due to their rugged terrains and remote locations?
    a. Nepal and Bhutan
    b. The Indian states of Punjab and Kerala
    c. Pakistan and Bangladesh
    d. Sri Lanka and Maldives
    e. The Indian states of Orissa and Jharkand

15. Why does Sri Lanka have the highest life expectancy and lowest under-age-5-mortality rates in South Asia?
    a. It has a large software development industry
    b. It has established universal primary education and inexpensive medical clinics
    c. It has received a great deal of foreign direct investment
    d. It is a very rich country
    e. It receives large amounts of foreign aid

Summary of South Asia

Total Population of South Asia: About 1.587 billion

Population Indicators for South Asia

	Highest (country and value)	Region Average	Lowest (country and value)
Population 2010 (millions)	India: 1,188.8	226.8	Maldives: 0.3
Density per sq km	Bangladesh: 1,142	475	Bhutan: 15
RNI	Pakistan: 2.3	1.7	Sri Lanka: 1.2
TFR	Pakistan: 4.0	2.9	Bangladesh; Sri Lanka: 2.4
Percent Urban	Maldives; Pakistan: 35%	27%	Sri Lanka: 15%
Percent < 15	Pakistan: 38%	32%	Sri Lanka: 26%
Percent > 65	Sri Lanka: 6%	5%	Bangladesh, Nepal, Pakistan: 4%
Net Migration (per 1000;2000-05)	Bhutan: 2.9	-0.5	Sri Lanka: -3.0

Development Indicators for South Asia

	Highest (country and value)	Region Average	Lowest (country and value)
GNI per capita/PPP 2008	Maldives: $5,290	$3,236	Nepal: $1,110
GDP Avg. Annual Growth (2000-2008)	India: 7.9%	5.6%	Nepal: 3.5%
Human Development Index (2007)	Maldives: 0.771	0.633	Bangladesh: 0.543
Percent Living below $2/day	Bangladesh: 81.3%	66.9%	Sri Lanka: 39.7%
Life Expectancy 2010	Sri Lanka: 74	68	India; Nepal: 64
< 5 Mortality 2008	Pakistan: 89 per 1000	56 per 1000	Sri Lanka: 15 per 1000
Gender Equity 2008	Bangladesh: 106	93	Pakistan: 80

Chapter 13

Southeast Asia

Learning Objectives

> ➢ This chapter introduces the 11 states of Southeast Asia
> ➢ You will learn about the region's long association with the global economy
> ➢ You will gain a deeper understanding of different types of agricultural systems and how these fit into globalization
> ➢ You will learn about the strategic location of Southeast Asia, and how this has influenced geopolitics
> ➢ The following terms and concepts will be developed: entrepôt, crony capitalism, swidden, and shifted cultivators

Chapter Outline

1. **Introduction:** Southeast Asia includes 11 countries, usually divided into two sub-regions: (1) mainland Southeast Asia, composed of Burma, Thailand, Cambodia, Laos, and Vietnam; and (2) insular—or maritime—Southeast Asia, composed of Indonesia, the Philippines, Malaysia, Singapore, Brunei, and East Timor; Malaysia actually occupied both mainland and insular Southeast Asia
   A. Southeast Asia occupies significant role in global economy; region includes some of world's most globally networked countries (e.g., Singapore)—but also some that are most resistant to globalization (e.g., Burma)
   B. Southeast Asia's involvement in large world has historic precedents—associated with Chinese and Indian cultures, traders from Southwest Asia, and Western colonialism
   C. Southeast Asia's resources and strategic location contributed to the region being a zone of political tension throughout twentieth-century
2. **Environmental Geography**
   A. **The Tragedy of the Karen:** The Karen, a tribal group, never fully incorporated into Burmese kingdom; during British colonial rule, the Karen territory became part of Burma—and Karen Christians obtained positions in Burma's colonial government; after independence, Karen lost favored position and conflict between Karen and Burmans ensued; the Karen rebelled and established insurgent state; Burmese army largely successful in defeating Karen separatists, in part, because of agreement made with Thailand; Thai government agreed to prevent Karen soldiers from finding sanctuary within its territory, in exchange for access to logging of Burma's teak forests
   B. **The Deforestation of Southeast Asia:** deforestation and related environmental problems are major issues throughout region; globalization has had profound effect—notably export-oriented logging; colonial powers cut forests for tropical hardwoods and naval supplies, and indigenous peoples cleared sections for agriculture; but rampant deforestation in late 1990s associated with large-scale international commercial logging
      a. **Causes of deforestation:** important to stress that in Southeast Asia, agriculture and population growth are *not* main causes of deforestation—most are cleared so that the wood products can be exported to other (usually industrialized) countries; also, forests cleared for establishment of plantations
      b. **Local Patterns of Deforestation**
         i. **Malaysia:** long-time leading exporter of tropical hardwoods; logging extensive in states of Sarawak and Sabah on island of Borneo

      ii. **Indonesia:** contains two-thirds of region's forest areas and 10 percent of world's true tropical rain forest; forests are deeply threatened—most of Sumatra's primary forests have been cleared and those in Kalimantan rapidly being cleared

      iii. **Thailand:** lost more than 50 percent of its forests between 1960 and 1980; logging bans in 1990s restricted commercial forestry; reforestation centers on planting of Australian eucalyptus trees—a nonnative species that cannot support local wildlife

      iv. **Other parts:** coastal areas losing mangrove forests; destruction threatens local fishing industries

  c. **Protected Areas:** Indonesia especially has created national parks and protected areas; other states also have created protected areas—attempts to preserve unique plants and wildlife—the orangutan, for example, lives only in small portion of northern Sumatra and island of Borneo

C. **Fires, Smoke, and Air Pollution:** commercial forest cutting—not small-scale agriculture—responsible for most burning; throughout 1990s, wildfires associated with logging (coupled with severe drought) led to massive wildfires—causing air pollution and other problems; efforts also to protect air quality hampered by industrial development and increase in vehicular traffic

D. **Patterns of Physical Geography:** insular Southeast Asia is one of world's three main zones of tropical rain forest; mainland Southeast Asia is located in tropical wet-and-dry zone

  a. **Mainland Environments:** rugged uplands interspersed with broad lowlands and deltas associated with large rivers; northern boundary lies in cluster of mountains connected to highlands of eastern Tibet and south-central China

      i. **Large Rivers:** several large rivers flow southward out of Tibet into Southeast Asia: Mekong, Irrawaddy, Red River, and Chao Phraya

      ii. **Khorat Plateau:** low, sandstone plateau in northeastern Thailand; water shortages and periodic droughts plague the area

  b. **Monsoon Climates:** Mainland Southeast Asia affected by seasonal winds known as the monsoon; hot and rainy season from May to October, warmer and dry season between November and April; two tropical climate regions in mainland—along coasts and climates is tropical monsoon climate with higher, more consistent rainfall; and rest of mainland Southeast Asia typified by tropical savanna climate—rainfall about half as much

  c. **Insular Environments:** island environment—Indonesia composed of about 13,000 islands; Philippines composed of about 7,000 islands

      i. **Indonesia:** dominated by four large islands of Sumatra, Borneo, Java, and Sulawesi; other important regional groupings of islands

      ii. **Philippines:** two largest and most important islands—Luzon in the north, and Mindanao in the south; in between is regional grouping known as the Visayan Islands

      iii. **Sunda Shelf:** world's largest expanse of shallow seas—an extension of the continental shelf extending from mainland through Java Sea between Java and Borneo; waters generally less than 200 feet deep

      iv. **Geology:** insular Southeast Asia geologically less stable; four tectonic plates converge: Pacific, Philippine, Indo-Australian, and Eurasian; earthquakes are frequent; volcanoes ever-present danger; tsunamis; typhoons

  d. **Island Climates:** more varied than mainland—depends on prevailing wind conditions of monsoon and topography; Indonesia, Singapore, Malaysia, and Brunei heavily influenced by equatorial location—consistently high temperatures year round; high and even distributed rainfall; typhoons frequent, especially from August to October—the Philippines is particularly prone

E. **Global Warming in Southeast Asia:** most people live in coastal or delta environments—hence vulnerable to sea-level rise; flooding already problem in low-lying cities—and will increase; agriculture (especially rice) might be adversely affected; rising sea-level might result in thousands of islands becoming submerged

  a. **El Niño:** climate change might intensify El Niño, resulting in more extreme droughts

    b. **Kyoto Accord:** all Southeast Asian countries ratified the 1997 Kyoto Accord—but since all are "developing" countries, none are obligated to reduce greenhouse gas emissions; overall emissions by global standards are relatively low, but deforestation is significant contributor to global climate change

3. **Population and Settlement:** Not heavily populated—compared to East or South Asia; most rugged mountainous areas are thinly inhabited; however, relatively dense populations found in deltas, coastal areas, and areas of fertile volcanic soil; demographic growth and family planning important concerns for most countries

    A. **Settlement and Agriculture:** much of insular Southeast Asia has infertile soil—unable to support intensive agriculture and high rural population densities; agriculture must be carefully adapted to land; Java is notable exception—has usually rich soils and agriculture is highly productive; demographic patterns of mainland straightforward: population is concentrated in agriculturally intensive valleys and deltas of large rivers—clusters around Chao Phraya River, Irrawaddy River, Red River Delta, and Mekong Delta; middle reaches of Mekong River less settled; in Cambodia, population concentrated around a large lake known as Tonle Sap

    B. **Swidden in the Uplands:** also known as shifting cultivation or "slash-and-burn" agriculture; swidden is practiced on both mainland and insular Southeast Asia

        a. **Swidden System:** small plots of forest are cleared (slashed); fallen vegetation is burned to transfer nutrients to soil; subsistence crops planted; yield remains high for several years until soil is exhausted; plots abandoned and allowed to revert to woody vegetation; farmer moves to new plot; cycle of cutting, burning, planting is termed *shifting cultivation*

            i. **Sustainability:** when population densities remain low, and sufficient territory is available—swidden is sustainable agricultural practice; but with increased populations, and commercial logging, rotation period is shortened and undercuts soil resources

        b. **Cash Crops:** when swidden is not possible, other crops cultivated; in northern Southeast Asian area known as *Golden Triangle*, main cash crop is opium, which is part of global drug trade

    C. **Plantation Agriculture:** region became center of plantation agriculture during European colonial period; rice, rubber especially; plantations still important: most of world's natural rubber is grown in Malaysia, Indonesia, and Thailand; sugarcane a major crop in Philippines and Indonesia; Indonesia major producer of tea; Malaysia and now Indonesia dominant in palm oil; coconut oil and copra in Philippines and Indonesia; Vietnam is world's second largest producer of coffee

    D. **Rice in the Lowlands:** lowland basins and deltas of mainland Southeast Asia largely devoted to intensive rice cultivation; Thailand currently world's largest rice exporter; in 2008, Thailand spearheaded creation of Organization of Rice Exporting Countries (OREC), which includes Cambodia, Laos, Burma, and Vietnam; in areas without irrigation, such as Khorat Plateau, rice yields are lower

    E. **Recent Demographic Change:** most countries have seen sharp decrease in birthrates

        a. **Population Contrasts:** Philippines has relatively high growth rate; Laos and Cambodia have relatively high rates; Thailand has lowered total fertility rate from 5.4 (in 1970) to 1.8—Thai government has supported family planning for population and health reasons (high incidence of HIV/AIDs); Singapore has particularly low fertility; Indonesia exhibits dramatic decline in fertility—resultant from government family planning efforts

        b. **Growth and Migration:** Indonesia has long history of transmigration program—relocation of people from densely settled islands (e.g., Java and Madura) to outer islands, especially East Kalimantan; social and environmental costs associated with transmigration; Philippines also forwarded relocation of migrants from Luzon to Mindanao—political and religious tensions accompanied movement

    F. **Urban Settlement:** not heavily urbanized, but urbanization rate is increasing; several Southeast Asian states have *primate cities*—single, large urban settlements that overshadow all others: Bangkok (Thailand) and Manila (the Philippines); urban primacy less pronounced in Vietnam (with both Hanoi and Ho Chi Minh City)

      a. **Secondary Cities:** Indonesia, Thailand, and the Philippines attempted to reduce over-urbanization in primate city through support of secondary, peripheral cities

4. **Cultural Coherence and Diversity:** Southeast Asia lacks historical dominance of single civilization; meeting ground for cultural traditions from South Asia, China, the Middle East, Europe, and North America

   A. **Introduction and Spread of Major Cultural Traditions:** contemporary cultural diversity is embedded in historical influences connected to major religions of Hinduism, Buddhism, Islam, and Christianity

      a. **South Asian Influences:** Hindu influence around 2,000 years ago; Hindu kingdoms established in coastal locations in Burma, Thailand, Cambodia, southern Vietnam, Malaysia, western Indonesia; Hinduism persists mostly on Indonesian islands of Bali and Lombok; Theravada Buddhism—also from South Asia—reached mainland Southeast Asia in thirteenth century—most inhabitants of lowland Burma, Thailand, Laos, and Cambodia are followers of Theravada Buddhism

      b. **Chinese Influences:** most pronounced in Vietnam, which was once province of China; South Asian influence notably lacking in Vietnam; East Asian (Chinese) cultural influences throughout rest of Southeast Asia linked to immigration from southern China; most urban areas retain large and cohesive Chinese communities; relationships between Chinese minorities and indigenous minorities strained in most locations

      c. **Arrival of Islam:** Muslim merchants from India and Southwest Asia arrived in Southeast Asia more than 1,000 years ago; by 1200s, religion spread from northern Sumatra through Malay Peninsula, Indonesian islands, and southern Philippines

         i. **Indonesia:** world's most populous Muslim country; exhibits significant internal diversity

         ii. **Malaysia:** Muslim state; Islamic fundamentalism is growing, but many Malay Muslims accuse fundamentalists of advocating social practices that are not religious in orientation; religious tensions continue (also in Indonesia)

         iii. **Philippines:** Islam diffused to southern Philippines by 1300s; northward spread stopped by arrival of Spanish colonists and conversion of much of archipelago to Christianity; the Philippines (but also East Timor) is only predominantly Christian country in Asia

      d. **Christianity and Indigenous Cultures:** Christian missionaries in late nineteenth and early twentieth centuries; many Vietnamese converted—but few elsewhere; missionaries most successful in highland areas; some modern tribal groups retain animist beliefs, but many have converted to Christianity; Indonesia experienced pronounced religious tension between Muslims and Christians in recent years

      e. **Religion and Communism:** Vietnam, Cambodia, and Laos have been communist states since 1975; religion was discouraged; Vietnam experiencing revival of faith

   B. **Geography of Language and Ethnicity:** language expresses long history of human movement; five major linguistic families present in region

      a. **The Austronesian Languages:** one of world's most widespread language family—from Madagascar to Easter Island; originated in Taiwan and spread with seafaring people who migrated from island to island; most of insular Southeast Asia dominated by this language family; tremendous diversity; in many areas, Malay has emerged as *lingua franca*, or common trade language

         i. **Philippines:** dominant language is *Filipino* (or *Pilipino*)—a form of standardized and modernized Tagalog, which is part of Austronesian language family; despite 350 years of Spanish colonialism, Spanish as language never widespread or unifying force

      b. **Tibeto-Burman Languages:** Burmese, spoken in Burma

      c. **Tai-Kadai Languages:** originate in southern China, now found throughout Thailand and Laos, parts of Burma; both Thai and Lao are national languages of Thailand and Laos, respectively

      d. **Mon-Khmer Languages:** 1,500 years ago, covered most of mainland Southeast Asia; contains Vietnamese and Khmer (national language of Cambodia); minor languages spoken by hill peoples; Vietnamese written with Chinese characters until French imposed Roman alphabet; Khmer written in own script, derived from India

    e. **Mainland Southeast Asia:** key point: national languages usually limited to core areas of densely populated lowlands, whereas peripheral uplands inhabited by tribal peoples speaking separate languages

  C. **Southeast Asian Culture in Global Context:** imposition of European colonial rule ushered in period of globalization; after decolonization, some countries (notably Burma) attempted to isolate from global system; others more receptive, including the Philippines and Thailand

    a. **Challenges:** cultural globalization viewed as threatening by some, especially within Islamic societies of Malaysia and Indonesia; Singapore likewise critical of some Western influences; English as "necessary evil"

5. **Geopolitical Framework:** Association of Southeast Asian Nations (ASEAN), composed of 10 states (excluding East Timor, which may gain admission by 2012); provides regional coherence

  A. **Before European Colonialism:** modern countries of mainland Southeast Asia mostly former indigenous kingdoms; significant kingdoms existed throughout present-day Burma, Cambodia, Vietnam, and Thailand; within insular Southeast Asia, numerous kingdoms throughout Malay Peninsula and islands of Sumatra and Java—but none territorially stable; no insular Southeast Asian kingdom led directly to modern state

  B. **Colonial Era:** Portuguese arrival in 1511 at Malacca in present-day Malaysia; Spanish arrival in 1500s (restricted to the Philippines); Dutch arrival by 1600s

    a. **Dutch Power:** Netherlands dominate throughout much of insular region, especially present-day Indonesia

    b. **British, French, and U.S. Expansion:** British concentrated on sea-lanes in Southeast Asia to facilitate empire in South Asia; established fortified posts along Strait of Malacca, including Singapore; Britain also occupied present-day Burma; French occupied present-day Vietnam, Laos, and Cambodia; the United States acquired the Philippines after defeating Spanish (and nationalist Filipinos who fought for independence)

  C. **Growing Nationalism:** organized resistance to European rule began in 1920s in mainland countries; agitation for independence intensified after World War II; Britain withdrew from Burma; Netherlands forced out of Indonesia; United States granted the Philippines its independence in 1946

  D. **The Vietnam War and Its Aftermath:** France determined to maintain colonies in Southeast Asia, led to war with Vietnamese from 1946 to 1954; in 1954, Vietnam was divided into two countries: North Vietnam (allied with China and the Soviet Union) and South Vietnam (allied with the United States); communist forces in both Laos (Pathet Lao) and in Cambodia (Khmer Rouge)

    a. **U.S. Intervention:** U.S. military advisors and politicians accepted *domino theory*—idea that if Vietnam fell to communism, so would Laos and Cambodia, and then neighboring countries of Burma, Thailand, and so on; increased U.S. military operations in Vietnam; by early 1970s, negotiated settlement of war concluded

    b. **Communist Victory:** withdrawal of U.S. forces allowed easy communist victories in Vietnam, Cambodia, and Laos; Vietnam reunited as single, communist-led country; Cambodia endured genocide at hands of Khmer Rouge—followed by insurgencies until United Nations peace settlement; corruption remains widespread and democratic institutions remain weak; in Laos, many Hmong and other tribal groups fled as refugees to Thailand and United States

  E. **Geopolitical Tensions in Contemporary Southeast Asia:** most serious geopolitical problems are internal rather than between countries; tensions also result from demands by tribal groups

    a. **Conflicts in Indonesia:** demands for greater autonomy or independence widespread throughout archipelago; notable examples include movements in western New Guinea; East Timor (which was successful in bid for independence); central and northern Sumatra; northern Sulawesi

b. **Regional Tensions in the Philippines:** Muslim separatist movements have long history in southern Philippines; some regional autonomy has been granted; emergence of more radical factions since late 1990s; U.S. military presence in response to global war on terror; other political problems in Philippines includes long-standing communist rebellion

c. **Quagmire of Burma:** war-ravaged country; wars through 1970s, 1980s, and 1990s pitted central government (dominated by Burmans) against country's varied non-Burman ethnic groups; democratic opposition against government meets with severe repression—Nobel Peace Prize winner Aung San Suu Kyi has been under house arrest for many years; rebelling ethnic groups seek to maintain authority over cultural traditions, lands, and resources—some insurgencies financed by drug trade

d. **Trouble in Thailand:** relatively peaceful, but significant anti-government protests from 2006 led to bloodless coup; ethnic and religious tensions more pronounced in Muslim dominated southern region; Thailand's conflict with Muslims has led to tensions with neighboring Malaysia

F. **International Dimensions of Southeast Asian Geopolitics:** radical Islamist groups pose greatest challenge

a. **Territorial Conflicts:** previously, several countries disputed boundaries; Philippines challenged Malaysia's control of Sabah on island of Borneo; dispute over Spratly Islands in South China Sea is ongoing—challenges by the Philippines, Malaysia, Vietnam, Brunei, China, and Taiwan

b. **ASEAN and Global Geopolitics:** Initially, ASEAN was alliance of non-socialist states; now all but East Timor are members; one purpose of ASEAN is to prevent any country (including the United States) from gaining undue influence in the region; ultimate international policy is to encourage conversation and negotiation over confrontation, and to enhance trade

c. **Global Terrorism and International Relations:** rise of radical Islamic fundamentalism has been greatest threat; notable group, Jemaah Islamiya (JI), which calls for creation of single Islamic states across Indonesia, Malaysia, southern Thailand, and southern Philippines; Indonesian forces have apparently neutralized JI; radical Islamist groups in southern Philippines and southern Thailand remain violent and militarily capable; many inhabitants of Southeast Asia (especially in Indonesia and Malaysia) remain wary of U.S. interventions as part of war on terror

6. **Economic and Social Development:** prior to late 1990s, most Southeast Asian states doing economically well; large foreign investment; but crises since 1997 have been followed by roller-coaster economies

A. **Uneven Economic Development:** region of strikingly uneven economic and social development; Indonesia has experienced booms and busts; Burma, Laos, and East Timor missed out of economic expansion; oil-rich Brunei and high-tech Singapore are among world's most prosperous countries

B. **The Philippine Decline:** 50 years ago was most highly developed Southeast Asian country; years of political corruption and mismanagement led to current failures of Philippines economy; during dictatorial presidency of Ferdinand Marcos (1968–1986) Philippines dominated by *crony capitalism*—friends of president were granted huge sectors of economy while those perceived as enemies had properties confiscated; Indonesia and Thailand also marked by crony capitalism; Philippines' economy supported largely by remittances from international migrants; by 21st century, Philippine economy showing signs of improvement, especially around former U.S. naval base of Subic Bay

C. **Regional Hub: Singapore:** transformed from *entrepôt* port city (a place where goods are imported, stored, and then transshipped) to one of the world's most prosperous and modern states; thriving high-tech manufacturing center and communications and financial hub of Southeast Asia; government has played a strong role in economy, but is considered somewhat repressive and undemocratic; authoritarian form of capitalism confronts new challenge from Internet—government worried about free expression it allows

D. **Malaysia's Insecure Boom:** experienced rapid economic growth over past several decades; development initially centered on natural resource extraction (tropical hardwoods and plantation products); more recently, manufacturing (especially high-tech); economy is multinational, but economy hard-hit since late 1990s; economic geography within Malaysia shows regional variation—most industrial development on west side of peninsular Malaysia; remainder of country dependent on agriculture and resource extraction; industrial wealth also concentrated in hands of ethnic Chinese

E. **Thailand's Ups and Downs:** rapid growth through 1980s and 1990s; collapsed in late 1990s; economic expansion uneven—Chao Phraya lowland and Bangkok have benefitted; northern area profiting from tourism; other regions, especially Khorat Plateau, are considerably more impoverished

F. **Indonesian Economic Development:** from poor country at time of independence (1949), to economically vibrant in 1970s, especially because of oil industry and commercial forestry; attracted multinational firms because of low wage labor; despite rapid growth, country remains poor—and remains dependent on unsustainable extraction of natural resources; pronounced geographic disparities—far eastern Indonesia has experienced little economic or social development; better conditions on Java and Madura

G. **Divergent Economic Paths: Vietnam, Laos, and Cambodia:** Vietnam's economy has prospered since moving from strict communist rule—now open to foreign investment and firms— but ongoing problems with upland tribal peoples; Laos and Cambodia continue to struggle, largely because of political instability and inadequate infrastructure; Laos also hindered by rough terrain and isolation (land-locked country); both countries are largely agricultural in orientation, and rely also on environmentally destructive logging and mining operations; Laotian government hoping to capitalize on export of hydroelectricity; Cambodia has experienced slight boom resultant from expanding textile sector, tourism, and mining; rampant problem of "land-grabbing" by political elites

H. **Burma's Troubled Economy:** one of region's most impoverished countries; has potential because of abundant natural resources (natural gas, oil, other minerals, water, timber) and relatively large expanse of fertile farmland; has political problems and refusal to engage with global economy; in early 1990s, began to open country slightly to foreign trade—efforts stalled by 2000; trade partners largely limited to China and India

I. **Globalization and the Southeast Asian Economy:** rapid but uneven integration into global economy; globalized industrial production remains controversial; movements to improve working conditions of laborers in export industries have emerged

J. **Issues of Social Development:** wide variation; Singapore ranks among world's leaders in regard to health and education; oil-rich Brunei, East Timor, Laos, and Cambodia rank near bottom; other countries have achieved relatively high levels of social welfare; most governments place high priority on basic education

Summary

- Some of the most serious problems created by globalization in Southeast Asia are environmental; deforestation as a result of commercial logging stands out
- Deforestation is also linked to domestic population and changes in settlement pattern
- Southeast Asia is characterized by tremendous cultural diversity; conflicts over language and religion are apparent in many countries
- The relative success of ASEAN has not solved political tension; many countries continue to dispute geographical, political, and economic issues; insurgencies remain strong in many countries
- Economic successes of ASEAN have been more limited; most trade is still oriented outward, rather than within region

Research or Term Paper Ideas

- Between 1975 and 1979, approximately one-quarter of Cambodia's eight million people died as a result of genocidal practices initiated by the Khmer Rouge. Conduct research on the genocide, but especially on the years following the genocide. How does the legacy of the genocide continue to influence Cambodia's society?

- Choose one of the many ethnic conflicts that current exists in Southeast Asia—such as the demands for autonomy among Muslim Filipinos, or the tensions between the upland hill peoples of Central Vietnam and the Vietnamese government. What is the basis of the conflict? When did it originate? What is the potential for a peaceful and satisfactory resolution to the problem?

- Learn more about the establishment of protected areas throughout Southeast Asia. How did these conservation efforts come about? Are they are any conflicts between these conservation strategies and the ability of rural peasants to maintain their way of life?

- East Timor is the newest sovereign state in Southeast Asia, having achieved independence in 2002. Learn more about the long struggle for East Timorese independence. How has an independent East Timor been integrated into the global economy?

Practice Quizzes

Answers appear at the end of this book

Vocabulary Matching: Match the term to its definition.

A. Animism
B. ASEAN
C. Bumiputra
D. Copra
E. Crony capitalism
F. Domino theory
G. Entrepôt
H. Golden Triangle

I. Khmer Rouge
J. Lingua franca
K. Ramayana
L. Shifted cultivators
M. Swidden
N. Transmigration
O. Tsunami
P. Typhoon

1. _____   Literally "Red (or communist) Cambodians," this left-wing insurgent group led by French-educated Marxists rebelled against the royal Cambodian government in the early 1960s and again in a peasants' revolt in 1967.

2. _____   An agreed-upon common language to facilitate communication on specific topics such as international business, politics, sports, or entertainment.

3. _____   A wide variety of tribal religions that generally focuses on nature's spirits and human ancestors.

4. _____   Literally "sons of the soil," the name given to native Malay, who are given preference for jobs and schooling by the Malaysian government.

5. _____   One of the two main epic poems of the Hindu religion.

6. _____   Very large sea waves induced by earthquakes.

7. _____   Migrants, with or without agricultural experience, who are transplanted by government relocation schemes.

8. _____   An international organization linking together the ten most important countries of Southeast Asian.

9. _____   The name given to tropical hurricanes in the western Pacific.

10. _____   The relocation of a nation's population from one location to another within its national territory.

11. _____   A U.S. geopolitical policy of the 1960s and 1970s that stemmed from the assumption that if Vietnam fell to the communists, the rest of Southeast Asia would soon follow.

12. _____   An area of northern Thailand, Burma, and Laos that is known as a major source region for heroin and is plugged into the global drug trade.

13. _____ A system in which close friends of a political leader are either legally or illegally given business advantages in return for their political support.

14. _____ A city and port that specializes in the transshipment of goods.

15. _____ Also called "shifting cultivation," or "slash-and-burn agriculture"; a form of cultivation in which forested or brushy plots are cleared of vegetation, burned, then planted to crops, only to be abandoned a few years later as soil fertility declines.

Multiple Choice: *Choose the word or phrase that best answers the question.*

1. Which phrase best describes the physical geography of Southeast Asia?
   a. Deserts and arid lands
   b. Extensive plains and grasslands
   c. Rolling hills and temperate forests
   d. Rugged terrain on the mainland and thousands of islands
   e. A combination of all of the above

2. In December 2004, which of the following environmental hazards caused tremendous loss of life and property damage in Southeast Asia?
   a. Volcanic eruption
   b. Typhoon
   c. Tsunami
   d. Forest fires
   e. Earthquake

3. What type of environmental damage is most prevalent in Southeast Asia?
   a. Deforestation
   b. Desertification
   c. Desiccation of lakes
   d. Improper disposal of nuclear waste
   e. Salinization of farmland from irrigation

4. Which climate type dominates in Southeast Asia?
   a. Continental
   b. Dry
   c. Highland
   d. Maritime
   e. Tropical Humid

5. What type of agriculture is most common in the uplands Southeast Asia?
   a. Swidden
   b. Rice cultivation
   c. Plantation agriculture
   d. Opium cultivation
   e. Pastoralism

6.   Which countries of Southeast Asia are below replacement level, based on their TFR (total fertility rate)?
      a.   Brunei, Singapore, and Thailand
      b.   Burma and Laos
      c.   Philippines and East Timor
      d.   Cambodia and Vietnam
      e.   Malaysia and Indonesia

7.   Which of the following groups is one of the foreign influences on Southeast Asia?
      a.   European colonists
      b.   Immigrants from China and South Asia
      c.   Muslim merchants
      d.   A and C above
      e.   A, B, and C above

8.   Which country of Southeast Asia had its culture most profoundly chanced by European colonization?
      a.   Vietnam, by the French
      b.   Philippines, by the Spanish
      c.   Malaysia, by Britain
      d.   Indonesia, by the Netherlands
      e.   Burma, by the British

9.   What attracted the earliest Europeans (the Portuguese) to Southeast Asia?
      a.   Sugarcane and rice
      b.   Gold
      c.   Nutmeg and cloves
      d.   Oil
      e.   Cotton

10.   What European country became the dominant colonial power in Southeast Asia?
      a.   Britain
      b.   France
      c.   Netherlands
      d.   Portugal
      e.   Spain

11.   Which country of Southeast Asia was a colony of the United States from 1898 until 1946?
      a.   Vietnam
      b.   Philippines
      c.   Laos
      d.   Cambodia
      e.   Indonesia

12.   What was the name of the theory that held that if Vietnam fell to the communists, so would its neighbors, Laos and Cambodia, then Burma, Thailand, and perhaps all of Southeast Asia?
      a.   Weakest link theory
      b.   Manifest destiny
      c.   Domino theory
      d.   Diffusion theory
      e.   Contagion theory

13. What is ASEAN's core mission?
    a. To regulate trade among the countries of Southeast Asia
    b. To develop environmental standards for member countries
    c. To prevent countries from outside Southeast Asia from gaining undue influence in the region
    d. A and B above
    e. A, B, and C above

14. What Southeast Asian countries have had the greatest developmental successes in the region?
    a. Brunei and Indonesia
    b. Laos and Cambodia
    c. Philippines and Burma
    d. Singapore and Malaysia
    e. Vietnam and Thailand

15. What is the nature of the correlation between economic and social indicators in Southeast Asia?
    a. Countries with favorable economic indicators tend to have favorable social indicators
    b. Countries with favorable economic indicators have less favorable social indicators
    c. On the Mainland, favorable economic indicators are correlated with favorable social indicators, but throughout Insular Southeast Asia, favorable economic indicators are correlated with unfavorable social indicators
    d. Throughout Insular Southeast Asia, favorable economic indicators are correlated with favorable social indicators, while on Mainland Southeast Asia, favorable economic indicators are correlated with unfavorable economic indicators
    e. There is no correlation between social and economic indicators

Summary of Southeast Asia

Total Population of Southeast Asia: Almost 600 million

Population Indicators for Southeast Asia

	Highest (country and value)	Region Average	Lowest (country and value)
Population 2010 (millions)	Indonesia: 235.5	54.3	Brunei: 0.4
Density per sq km	Singapore: 7,526	798	Laos: 27
RNI	East Timor: 3.1	1.5	Singapore; Thailand: 0.6
TFR	East Timor: 5.7	2.7	Singapore: 1.2
Percent Urban	Singapore: 100%	45%	Cambodia: 20%
Percent < 15	East Timor: 45%	30%	Singapore: 18%
Percent > 65	Singapore: 9%	5%	Burma (Myanmar); Brunei; Cambodia; East Timor: 3%
Net Migration (per 1000;2000-05)	Singapore: 22	1.8	Laos: -2.4

Development Indicators for Southeast Asia

	Highest (country and value)	Region Average	Lowest (country and value)
GNI per capita/PPP 2008	Brunei: $50,770	$13,898	Cambodia: $1,860
GDP Avg. Annual Growth (2000-2008)	Cambodia: 9.8%	5.9%	East Timor: 1.9%
Human Development Index (2007)	Singapore: 0.944	0.725	East Timor: 0.489
Percent Living below $2/day	Laos: 76.8%	38%	Singapore: <2%
Life Expectancy 2010	Singapore: 81	69	Burma (Myanmar): 58
< 5 Mortality 2008	Burma (Myanmar): 98 per 1000	45 per 1000	Singapore: 3 per 1000
Gender Equity 2008	Philippines: 102	95	Laos: 87

Chapter 14

Australia and Oceania

Learning Objectives

➤ This chapter introduces Australia, New Zealand, and the Oceania realm
➤ You will learn about the unique island environments of Oceania and the important cultural adaptations of people within this realm
➤ You will learn about the unique biogeographies associated with islands, and the negative impacts of introduced species
➤ You will learn how this far-flung realm is unevenly connected with the global economy
➤ The following terms and concepts will be introduced: atoll, archipelago, coral reefs, exotic species and extinctions, Aborigines

Chapter Outline

1. **Introduction:** this region includes the island continent of Australia, the country of New Zealand, and Oceania—a collection of islands that reach from New Guinea to the U.S. state of Hawaii
   A. The region has long been settled by humans; European colonization during the eighteenth century began the process of globalization
   B. Ongoing political and ethnic unrest typifies many parts of Oceania; Fiji provides one such illustration
   C. The region exhibits a diversity of environments, including the desert interior of Australia (known as the Outback); Oceania, however, is a tropical realm of islands, coral reefs, and atolls
   D. Oceania is divided into three broad sub-regions
      a. Melanesia (meaning "dark islands") contains the islands of New Guinea, the Solomon Islands, Vanuatu, and Fiji
      b. Polynesia ("many islands") is located east of Melanesia; this linguistically unified sub-region includes the French-controlled Tahiti in the Society Islands, the Hawaiian Islands, and smaller political states such as Tonga, Tuvalu, and Samoa; New Zealand is also considered part of Polynesia because its indigenous peoples, the *Maori*, share many cultural and physical characteristics with the other peoples of the sub-region
      c. Micronesia ("small islands") is a sub-region north of Melanesia and east of Polynesia; it includes the microstates of Nauru, the Marshall Islands, and the U.S. territory of Guam
2. **Environmental Geography:** the geology and climate of the Pacific Ocean define much of the physical geography of Oceania
   A. **Environments at Risk:** despite relatively small populations, many areas face significant human-induced environmental problems; also the region is at risk of natural events, including earthquakes, tropical cyclones, and (in Australia) droughts
   B. **Global Resource Pressures:** globalization of resource extraction has exacted environmental toll on region
      a. **Mining Operations:** international mining operations, especially throughout Australia, Papua New Guinea, New Caledonia, and Nauru have resulted in severe ecological damage
      b. **Deforestation:** the clearing of forests for pasture land has been devastating throughout Australia; islands within Oceania (e.g., Papua New Guinea) are also threatened from logging operations

    c. **Exotic Plants and Animals:** introduction of exotic (non-native) plants and animals have caused problems for endemic (native) species; non-native rabbits in Australia, for example, have adversely affected local ecosystems; introduction of brown tree snake in Guam has led to significant problems and both ecological and economical damage

C. **Global Warming in Oceania:** adverse effects of climate change already experienced in Oceania—mountain glaciers of New Zealand are retreating; Australia suffers from frequent droughts and wildfires; warmer ocean waters have caused damage to coral reefs; rising sea levels are flooding several low-lying island nations; projections for the future include stronger tropical cyclones and additional coastal degradation

    a. **Actions and Policies:** until 2007 Australia opposed to Kyoto Protocol—in part because Australia is the world's largest exporter of coal; current government supports regulations, but is opposed by coal industry, aluminum manufacturers, and agriculture; many Pacific nations, including Tuvalu, Kiribati, and the Marshall Islands, have formed a political union lobbying for global solution to climate change

D. **Australian and New Zealand Environments**

    a. **Landform Regions:** three major landform regions dominate Australia: Western Plateau; interior lowlands (the Outback); and the Eastern Highlands; offshore is the Great Barrier Reef; New Zealand, composed of two main islands, is volcanic in origin

    b. **Climate:** zones of higher precipitation encircle Australia's arid center; parts of southern and southwestern corner of Australia are dominated by Mediterranean climate and distinctive vegetation, known as *mallee*, a scrubby eucalyptus woodland; climates in New Zealand influenced by latitude, moderating effects of the Pacific Ocean, and topography; North Island is distinctive sub-tropical; southern region is cooler

    c. **Island Climates:** many Pacific islands receive abundant precipitation

E. **The Oceanic Realm**

    a. **Creating Island Landforms:** Much of Melanesia and Polynesia is part of seismically active Pacific Basin; volcanic eruptions, earthquakes, and tsunamis (seismically induced sea waves) are common; most islands are oceanic, having originated from volcanic activity on the ocean floor; larger active and recently active islands are termed *high islands*—these often rise to considerable heights and cover large areas; the island of Hawaii, the largest of the Hawaiian Island chain, is an example; the Hawaiian Islands, also a geological *hot spot*, where slowly moving oceanic crust passes over a vast supply of magma from Earth's interior, thus creating a chain of volcanic islands; other examples of high islands include those of French Polynesia, including Bora Bora

        a. **Coral Reefs:** in tropical latitudes, most high islands are ringed by coral reefs—these grow in shallow waters near shore

        b. **Atoll:** the combination of narrow sandy islands, barrier coral reefs, and shallow central lagoons is known as an *atoll*; these are characteristically circular or oval in shape but have many variations; Polynesia, Melanesia, and Micronesia are dotted with extensive atoll systems

3. **Population and Settlement:** long history of indigenous and European settlement; in Australia and New Zealand, Anglo-European migration structured the distribution and concentration of contemporary populations; in Oceania, population geographies determined by needs of native peoples

A. **Contemporary Population Patterns:** Australia is highly urbanized; people concentrated mostly in sub-tropical south and east; low population densities in the interior—mostly inhabited by Aborigines (indigenous peoples); New Zealand—most inhabit northern island; Oceania overwhelmingly rural

B. **Historical Settlement:** region's remoteness from world's early population centers; peripheral to dominant migratory paths of earlier peoples

    a. **Peopling the Pacific:** large islands of New Guinea and Australia, proximate to Asian landmass, first settled—about 40,000 years ago ancestors of today's Aboriginal (indigenous) populations settled; Eastern Melanesia settled later—about 3,500 years ago; Oceania settled by sea-faring peoples; continued movement into western Polynesia and by 800 CE reached New Zealand, Hawaiian Islands, and Easter Island; population pressures might have reached crisis stage on small islands, thus impelling subsequent moves to colonize other islands

- b. **European Colonization:** Dutch colonial activity in seventeenth century, followed by British colonial activities in eighteenth century; large-scale European settlement emerges with British establishment of Australia as a penal colony; new settlers conflict with Aborigines, leading to death of many Aborigines; also removal of Aborigines from lands; New Zealand attracted British settlers; similar to Aborigines, the indigenous peoples of New Zealand—the Maori—lost control of their lands; the native Hawaiians also lost their lands, especially after 1893 with the United States' overthrow of the Hawaiian monarchy and formal political annexation in 1898

C. **Settlement Landscapes:** presents mixture of local and global influences
- a. **The Urban Transformation:** both Australia and New Zealand are highly urbanized; vast majority of populations live in city and suburban environments; other urban areas in Oceania include Port Moresby in Papua New Guinea—this city, though, unlike those in Australia and New Zealand, presents many of same problems of urban underdevelopment found in less developed regions
- b. **The Rural Scene:** rural landscapes across Australia and Pacific express complex mosaic of cultural and economic influences; in some settings, indigenous ways of life still found, but these landscapes becoming rare; in Australia, sheep and cattle dominate rural landscape
- c. **New Zealand's Landscapes:** rural landscape includes variety of agricultural activities, include pastoral agriculture and dairy operations (mostly in lowlands of the north)
- d. **Rural Oceania:** on high islands with more water, denser populations take advantage of diverse agricultural opportunities; otherwise, fishing predominates; several types of rural settlements are apparent, including village-centered shifting cultivation (e.g., in rural New Guinea), which are often interspersed with subsistence crops and some commercial crops; elsewhere, traditional agricultural patterns are found; commercial plantations established on some islands, including the Solomon Islands and Vanuatu (copra, cocoa, and coffee) and Fiji and Hawaii (sugarcane)

D. **Diverse Demographic Paths**
- a. **High Population Growth Rates:** a problem on many islands, including Vanuatu and the Marshall Islands; exceptionally pressing problem on smaller island groups in Micronesia and Polynesia
- b. **Migration:** out-migration is high from several islands (e.g., Tonga and Samoa)—driven largely by unemployment; both Australia and New Zealand attract migrants; New Caledonia attracts migrants because of mining boom

4. **Cultural Coherence and Diversity:** worldwide processes of globalization have redefined region's cultural geography

A. **Multicultural Australia:** dominated by colonial European roots
- a. **Aboriginal Imprints:** for thousands of years, Australia dominated by Aborigines; hunting-and-gathering societies made little impact on landscape; populations decimated after arrival of British; indigenous cultures preserve and growing native peoples movement is apparent; however, most Aborigines live in cities and few practice traditional lifestyles—hence growing interest in preserving traditional cultural values, especially in the Outback, where languages remain strong and growing number of speakers; aspects of Aboriginal spiritualism being preserved
- b. **A Land of Immigrants:** Twentieth century dominated by European migration; in late nineteenth century, however, laborers recruited from the Solomons and New Hebrides—these Pacific Islanders (known as *kanakas*) were spatially and socially segregated from Anglo employers; Australia previously had "White Australia policy," which restricted non-Anglo-European migration; policy rescinded and now considerable migration from Asia; some tensions as result of new influx of migrants

   B. **Cultural Patterns in New Zealand:** parallels patterns in Australia; Maoris historically discriminated against—lost much of original homelands; also committed to preserving their way of life
   C. **Mosaic of Pacific Cultures:** in more isolated places, traditional cultures largely insulated from outside influences
       a. **Language Geography:** most of native languages of Oceania belong to Austronesian language family, which encompasses the Pacific, parts of Southeast Asia, and Madagascar; Malayo-Polynesian subfamily includes most of the related languages of Micronesia and Polynesia; Melanesia's language geography more complex and incompletely understood— more than 1,000 languages, for example, have been identified on the island of New Guinea
       b. **Village Life:** many different types of traditional patterns of social life; throughout Melanesia, dominance of small villages occupied by single clan or family; in Polynesia, also, village life—but often strong class-based relations apparent
       c. **External Cultural Influences:** most Pacific islands have witnessed tremendous cultural transformations in past 150 years: new settlers, values, technological innovations; European colonialism transformed the region; cultural makeup impacted through various immigration systems; Hawaii is notable for cultural mix
5. **Geopolitical Framework:** complex interplay of local, colonial-era, and global-scale forces
   A. **Roads to Independence:** region is marked by newness and fluidity of political boundaries; oldest independent states are Australia and New Zealand—both were twentieth century creations; most of the newly independent Pacific *microstates* retain political and economic ties to former colonial powers
       a. **Colonial Ties:** most independent states in Oceania achieved this status in 1970s; many retain legacy of colonial rule—former U.S. territories, for example, reflect past and current military presence: Bikini Atoll and other places used as nuclear test grounds; others remain as sites of U.S. military installations; France has been less willing to give up authority throughout Oceania
   B. **Persisting Geopolitical Tensions:** cultural diversity, colonial legacy, youthful states, and rapidly changing political map contribute to ongoing geopolitical tensions
       a. **Native Rights in Australia and New Zealand:** indigenous peoples in both countries have used political process to gain more control over land and resources; concessions made, including 1993 Native Title Bill in Australia, which compensated Aborigines for lands lost and provided right to gain title to unclaimed lands they still occupied; more recent efforts have met with resistance from Anglo-Australians; in New Zealand, Maori land claims also generate controversy
       b. **Ongoing Conflicts in Oceania:** Papua New Guinea contends with ethnic tensions; French colonial presence in Pacific continues to pose political problems
       c. **Influences from the Two Chinas:** Taiwan has long provided economic aid to island nations of Nauru, Tuvalu, the Solomon Islands, Marshall Islands, and Palua in exchange for political allegiance; the People's Republic of China, however, is emerging as major player in region
           a. **China's Objectives:** draw upon region's natural resources to support its own economy; and to neutralize (and replace) Taiwan's political and economic influence in the region
           b. **Australia and New Zealand:** both still play key political roles in South Pacific
6. **Economic and Social Development:** region reflects diversity of economic situations resulting in both wealth and poverty; overall economic future of Pacific realm remains variable because of small domestic markets, peripheral position in global economy, and diminishing resource base
   A. **The Australian and New Zealand Economies:** Australia's past economic wealth built on cheap extraction and export of raw materials; export-oriented agriculture a stalwart; mining sector very important—one of the world's mining superpowers; mining growing because of trade with China; expanding links with Asian markets offers promise for future; New Zealand traditionally reliant upon exports but now transformed into market-oriented country

B. **Oceania's Economic Diversity:** varied economic activities shape Pacific Island nations; much of life oriented around subsistence-based agriculture, including shifting cultivation and fishing, also prevalent are plantation economies, mining, and timber activities; some islands have capitalized on global tourism
   a. **Melanesia:** least developed and poorest of Oceania—benefitted less from tourism and subsidies from wealthy colonial and ex-colonial powers; other activities include coffee growing, cattle grazing, and tourism
      a. **Economic Impact of Mining:** mining economies dominate New Caledonia and Nauru
   b. **Micro-Polynesian Economies:** economic conditions depend on both local subsistence economies and economic linkages beyond region; some island groups receive substantial subsidies from France or the United States
C. **The Global Economic Setting:** many international trade flows link the region to other parts of the world; Australia and New Zealand dominate global trade patterns in the region; other ties come in form of capital investment, especially from the U.S. and Japan
   a. **Asia-Pacific Economic Cooperation Group (APEC):** organization designed to encourage economic development in Southeast Asia and the Pacific Basin
   b. **Closer Economic Relations (CER) Agreement:** signed in 1982, designed to promote economic integration between Australia and New Zealand
   c. **Other Nations of Oceania:** many are closely tied to external powers, including the United States and Japan; also benefit from proximity to Australia and New Zealand
D. **Continuing Social Challenges**
   a. **Australia and New Zealand:** generally high levels of social welfare; confronted with challenges found elsewhere in developed world, including cancer and heart disease; Australia's rate of skin cancer is among world's highest; both countries provide high-quality health care; social conditions of indigenous peoples less favorable
   b. **Oceania:** levels of social welfare generally higher than expected, given levels of poverty; however, many island nations have invested heavily in health and education services

Research of Term Paper Ideas

- Conduct library research to learn more about the treatment of Aborigines in Australia and the Maori in New Zealand. Compare the experiences of these two groups with that of Native Americans in North America.

- The Maoris have elaborate tattooing. Explore the history of Maori tattoos. What was the importance of the tattoos? How were they created? Is tattooing still widely practiced among the Maori?

- Australia and Oceania have had significant problems resulting from biological invasions. In Australia, for example, the introduction of non-native rabbits and toads has resulted in significant ecological damage—often at considerable economic expense. Learn more about invasive plants and animals. How are invasive plants and animals introduced to other ecosystems? How significant is the problem of invasive species—not only in Australia and Oceania, but in other parts of the world. How has globalization contributed to the increase in biological invasions?

Practice Quizzes

Answers appear at the end of this book

Vocabulary Matching: Match the term to its definition.

A. Aborigine
B. Archipelago
C. Atoll
D. Haoles
E. High islands
F. Hot spot
G. Kanaka
H. Maori

I. Melanesia
J. Micronesia
K. Native Title Bill
L. Oceania
M. Outback
N. Polynesia
O. Uncontacted peoples
P. Viticulture

1. _____   Larger, more elevated islands, often focused around recent volcanic activity.

2. _____   Pacific Ocean region that includes the culturally diverse, generally small islands, including the Mariana Islands, Marshall Islands, and others.

3. _____   Australia's huge, dry interior, as thinly settled as the Sahara Desert.

4. _____   A major world sub-region that includes New Zealand and the major islands of Melanesia, Micronesia, and Polynesia.

5. _____   An indigenous inhabitant of Australia.

6. _____   Pacific Ocean region, broadly unified by language and cultural traditions, includes the Hawaiian Islands, Marquesas Islands, Tuamotu Archipelago, American Samoa, Tonga, Kiribati, and others.

7. _____   Indigenous Polynesian people of New Zealand.

8. _____   Low, sandy islands made from coral, often oriented around a sandy lagoon; volcanic eruption is the first step in the creation of this landform.

9. _____   Island groups, often oriented in an elongated pattern.

10. _____   Melanesian workers imported to Australia, historically often concentrated along Queensland's "sugar coast."

11. _____   "Dark Islands" of New Guinea, Solomon Islands, Vanuatu, Fiji.

12. _____   A supply of magma that produces a chain of mid-ocean volcanoes atop a zone of moving oceanic crust.

13. _____   Ethnic and cultural groups that have yet to be "discovered" by the Western world.

14. _____   Light-skinned Europeans or U.S. citizens in the Hawaiian Islands.

15. ____ Law that gave Australia's aboriginal population greater control over their sacred lands.

Multiple Choice: *Choose the word or phrase that best answers the question.*

1. Why do seismic hazards, periodic droughts in Australia and violent tropical cyclones pose a greater threat now than they did in the past?
    a. Because of global warming, these events are becoming more frequent
    b. Shifting tectonic plates are causing the seismic events to become stronger, while global warming is causing the tropical cyclones to become more powerful and the droughts to last longer
    c. New settlements have made increasing populations vulnerable to these problems
    d. A and B above
    e. A, B, and C above

2. Which of the following poses a threat to the environment of Australia/Oceania?
    a. Logging by transnational corporations and clear-cutting forests to create pastures
    b. Major mining operations
    c. Fallout from nuclear testing
    d. A and B above
    e. A, B, and C above

3. Which of the following exotic species has wiped out nearly all the native bird species in Guam?
    a. Brown tree snake
    b. Rabbit
    c. Rat
    d. Sheep
    e. Pigs

4. What is the driest part of Australia/Oceania?
    a. Australia's center
    b. Fiji
    c. New Zealand's North Island
    d. Tasmania
    e. The western coastline of Vanuatu

5. Volcanic eruption is the first step in the creation of which of these features?
    a. Atoll
    b. High islands
    c. Tsunamis
    d. A and B above
    e. A, B, and C above

6. Where do the majority of the people of the region of Australia/Oceania live?
    a. Australia
    b. Hawaii
    c. New Zealand
    d. Marshall Islands
    e. Rural villages throughout the region

7. Which of these statements about Aborigines is/are correct?
    a. Aborigines originated in Southeast Asia and were hunters and gatherers when the first Europeans arrived
    b. In Tasmania, Aborigines were hunted down and killed
    c. Aborigines were run off their lands, and although their populations dwindled, they survived in the Australian Outback
    d. A and B
    e. A, B, and C

8. Which of these has been a major factor in the transformation of cities in Oceania in the past 50 years?
    a. Tourism
    b. Military conflict
    c. Urban planning departments
    d. Mining operations
    e. Plantation agriculture

9. Which of these places has experienced net out-migration between 2005 and 2010?
    a. Australia
    b. Tonga and Samoa
    c. French Polynesia and Guam
    d. New Caledonia and Vanuatu
    e. New Zealand

10. Which of the following statements about the Maori is/are correct?
    a. After declining with initial European contacts and conflicts, the Maori population began to rebound in the twentieth century
    b. Maori and Aborigines are members of the same ethnic group whose only difference is their country of residence
    c. The Maori are more numerically important and culturally visible in New Zealand than the Aboriginal counterparts in Australia
    d. A and B above
    e. A and C above

11. What is the major difference between the boundaries created by indigenous peoples of Australia/Oceania and those created by European colonists?
    a. Indigenous people naturally evolved clear ethnographic boundaries; Europeans replaced them with physiographic boundaries
    b. Indigenous peoples had established firm boundaries with their neighbors; Europeans eliminated these boundaries
    c. Indigenous peoples preferred nuanced, often fluid boundaries; Europeans preferred more precise, yet unstable borders
    d. Indigenous peoples preferred physiographic boundaries; Europeans preferred geometric boundaries
    e. Indigenous peoples used local landmarks to mark boundaries; Europeans used fences

12. Which foreign country continues to maintain a colonial presence in Australia/Oceania, causing tensions in New Caledonia?
    a. United States
    b. Portugal
    c. Netherlands
    d. France
    e. Britain

13. What is Aoteroa?
    a. The name of the official airline of Australia
    b. The name that the Maoris call New Zealand
    c. The indigenous peoples of Tasmania
    d. An indigenous, region-wide environmental movement in Australia/New Zealand
    e. A flightless bird with hairy feathers native to Australia

14. Which country of Australia/Oceania is grouped among the world's developed countries?
    a. Australia
    b. New Zealand
    c. French Polynesia
    d. A and B above
    e. A and C above

15. Which of these pairings of country and economic activity is/are correct?
    a. Hawaii – assembly plant industrialization
    b. New Caledonia – nickel mining
    c. Nauru – phosphate mining
    d. A and B
    e. B and C

Summary of Australia and Oceania

Total Population of Australia and Oceania: 43 million

Population Indicators for Australia and Oceania

	Highest (country and value)	Region Average	Lowest (country and value)
Population 2010 (millions)	Australia: 22.4	2.2	Nauru; Tuvalu: 0.01
Density per sq km	Tuvalu: 376	134	Australia: 3
RNI	Marshall Islands: 2.8	1.7	Palau: 0.6
TFR	Solomon Islands: 4.4	3.2	Australia: 1.9
Percent Urban	Nauru: 100%	52%	Papua/New Guinea: 13%
Percent < 15	Marshall Islands; Solomon Islands: 41%	33%	Australia: 19%
Percent > 65	Australia; New Zealand: 13%	5%	Nauru: 1%
Net Migration (per 1000;2000-05)	Australia: 4.8	-4.1	Samoa: -18.4

Development Indicators for Australia and Oceania

	Highest (country and value)	Region Average	Lowest (country and value)
GNI per capita/PPP 2008	Australia: $37,250	$8,968	Papua/New Guinea: $2,130
GDP Avg. Annual Growth (2000-2008)	Australia: 3.3%	3.1%	Papua/New Guinea: 2.9%
Human Development Index (2007)	Australia: 0.970	0.756	Papua/New Guinea: 0.541
Percent Living below $2/day	Papua/New Guinea: 58%	Too little data	Australia: <2%
Life Expectancy 2010	Australia: 81	69	Nauru: 56
< 5 Mortality 2008	Papua/New Guinea: 69 per 1000	35 per 1000	Australia; New Zealand: 9 per 1000
Gender Equity 2008	Kiribati: 107	101	Solomon Islands: 93

Answers to Practice Quizzes

Chapter 1: Diversity Amid Globalization

Vocabulary Matching:

1-L; 2-F; 3-C; 4-J; 5-M; 6-G; 7-O; 8-D; 9-E; 10-P; 11-B; 12-K; 13-A; 14-I; 15-H

Multiple Choice:

1-A; 2-B; 3-A; 4-D; 5-E; 6-E; 7-C; 8-B; 9-C; 10-D; 11-C; 12-A; 13-E; 14-D; 15-A

Chapter 2: The Changing Global Environment

Vocabulary Matching:

1-A; 2-L; 3-F; 4-H; 5-B; 6-G; 7-I; 8-C; 9-E; 10-K; 11-D; 12-J; 13-H; 14-O; 15-N

Multiple Choice:

1-E; 2-C; 3-A; 4-D; 5-E; 6-A; 7-A; 8-E; 9-E; 10-A; 11-E; 12-B; 13-D; 14-C; 15-A

Chapter 3: North America

Vocabulary Matching:

1-A; 2-M; 3-G; 4-P; 5-J; 6-F; 7-L; 8-B; 9-O; 10-K; 11-I; 12-E; 13-C; 14-D; 15-H

Multiple Choice:

1-D; 2-E; 3-D; 4-D; 5-C; 6-C; 7-A; 8-E; 9-E; 10-E; 11-C; 12-E; 13-E; 14-D; 15-D

Chapter 4: Latin America

Vocabulary Matching:

1-L; 2-D; 3-I; 4-B; 5-C; 6-F; 7-P; 8-G; 9-O; 10-K; 11-J; 12-M; 13-E; 14-A; 15-N

Multiple Choice:

1-D; 2-E; 3-D; 4-B; 5-E; 6-D; 7-C; 8-B; 9-D; 10-A; 11-D; 12-E; 13-C; 14-D; 15-B

Chapter 5: The Caribbean

Vocabulary Matching:

1-D; 2-G; 3-E; 4-J; 5-F; 6-L; 7-K; 8-H; 9-M; 10-O; 11-P; 12-B; 13-N; 14-C; 15-A

Multiple Choice:

1-B; 2-B; 3-C; 4-E; 5-D; 6-A; 7-E; 8-A; 9-B; 10-B; 11-E; 12-C; 13-E; 14-D; 15-A

Chapter 6: Sub-Saharan Africa

Vocabulary Matching:

1-D; 2-L; 3-O; 4-P; 5-J; 6-P; 7-A; 8-I; 9-B; 10-N; 11-F; 12-C; 13-G; 14-E; 15-M

Multiple Choice:

1-A; 2-B; 3-A; 4-E; 5-C; 6-E; 7-D; 8-E; 9-B; 10-C; 11-E; 12-D; 13-B; 14-C; 15-E

Chapter 7: Southwest Asia and North Africa

Vocabulary Matching:

1-D; 2-C; 3-P; 4-A; 5-M; 6-O; 7-E; 8-L; 9-N; 10-F; 11-B; 12-H; 13-K; 14-J; 15-I

Multiple Choice:

1-D; 2-D; 3-E; 4-E; 5-B; 6-E; 7-D; 8-E; 9-A; 10-C; 11-A; 12-B; 13-E; 14-A; 15-D

Chapter 8: Europe

Vocabulary Matching:

1-P; 2-G; 3-M; 4-L; 5-K; 6-J; 7-C; 8-F; 9-N; 10-E; 11-D; 12-A; 13-O; 14-I; 15-H

Multiple Choice:

1-B; 2-E; 3-E; 4-E; 5-C; 6-A; 7-E; 8-E; 9-A; 10-B; 11-C; 12-C; 13-A; 14-C; 15-C

Chapter 9: The Russian Domain

Vocabulary Matching:

1-B; 2-P; 3-M; 4-K; 5-H; 6-F; 7-O; 8-I; 9-J; 10-D; 11-A; 12-N; 13-C; 14-E; 15-L

Multiple Choice:

1-E; 2-A; 3-B; 4-E; 5-D; 6-E; 7-A; 8-C; 9-D; 10-D; 11-B; 12-C; 13-A; 14-E; 15-E

Chapter 10: Central Asia

Vocabulary Matching:

1-J; 2-K; 3-F; 4-B; 5-G; 6-A; 7-L; 8-E; 9-I; 10-D; 11-H; 12-C

Multiple Choice:

1-C; 2-D; 3-E; 4-C; 5-A; 6-E; 7-C; 8-E; 9-C; 10-A; 11-A; 12-A; 13-C; 14-B; 15-D

Chapter 11: East Asia

Vocabulary Matching:

1-G; 2-E; 3-P; 4-O; 5-A; 6-J; 7-M; 8-K; 9-F; 10-H; 11-D; 12-L; 13-C; 14-B; 15-I

Multiple Choice:

1-E; 2-B; 3-D; 4-D; 5-A; 6-A; 7-C; 8-B; 9-B; 10-A; 11-C; 12-E; 13-D; 14-E; 15-A

Chapter 12: South Asia

Vocabulary Matching:

1-G; 2-L; 3-D; 4-K; 5-O; 6-B; 7-I; 8-E; 9-C; 10-M; 11-J; 12-N; 13-F; 14-P; 15-A

Multiple Choice:

1-B; 2-E; 3-D; 4-D; 5-C; 6-A; 7-B; 8-D; 9-E; 10-C; 11-E; 12-D; 13-D; 14-A; 15-B

Chapter 13: Southeast Asia

Vocabulary Matching:

1-I; 2-J; 3-A; 4-C; 5-K; 6-O; 7-L; 8-B; 9-P; 10-N; 11-F; 12-H; 13-E; 14-G; 15-M

Multiple Choice:

1-D; 2-C; 3-A; 4-E; 5-A; 6-A; 7-E; 8-B; 9-C; 10-C; 11-B; 12-C; 13-C; 14-D; 15-A

Chapter 14: Australia and Oceania

Vocabulary Matching:

1-E; 2-J; 3-M; 4-l; 5-A; 6-N; 7-H; 8-C; 9-B; 10-G; 11-I; 12-F; 13-O; 14-D; 15-K

Multiple Choice:

1-C; 2-E; 3-A; 4-A; 5-D; 6-A; 7-E; 8-A; 9-B; 10-E; 11-C; 12-D; 13-B; 14-D; 15-E